TEXTBOOK OF PHYLOGENETICS

TEXTBOOK
OF
PHYLOGENETICS

By

Manju Yadav

Lecturer

Department of Zoology

M.M.H. College

Ghaziabad (U.P.)

(India)

DISCOVERY PUBLISHING HOUSE PVT. LTD.

NEW DELHI-110 002

Published by:

DISCOVERY PUBLISHING HOUSE PVT. LTD.
4383/4B, Ansari Road, Darya Ganj
New Delhi-110 002 (India)
Phone : +91-11-23279245; 23253475; 43596065
Mobile : +91 9811179893 / +91 9871656464
E-mail : discoverybooksindia@gmail.com
orderdphbooks@gmail.com
namitwasan9@gmail.com
web : www.discoverypublishinggroup.com

First Edition: **2010**

Reprinted: **2025**

ISBN: 978-81-8356-590-5

Textbook of Phylogenetics

Printed at:
Infinity Imaging Systems
Delhi (INDIA)

PREFACE

Evolution is regarded as a branching process, whereby populations are altered over time and may speciate into separate branches, hybridize together, or terminate by extinction. This may be visualized as a multi-dimensional character-space that a population moves through over time. The problem posed by phylogenetics is that genetic data are only available for the present, and fossil records (osteometric data) are sporadic and less reliable. Our knowledge of how evolution operates is used to reconstruct the full tree.

There are some terms that describe the nature of a grouping in such trees. For instance, all birds and reptiles are believed to have descended from a single common ancestor, so this taxonomic grouping is called monophyletic. "Modern reptile" is a grouping that contains a common ancestor, but does not contain all descendents of that ancestor (birds are excluded). This is an example of a paraphyletic group. A grouping such as warm-blooded animals would include only mammals and birds and is called polyphyletic because the members of this grouping do not include the most recent common ancestor. Thus, a phylogenetic tree is based on a hypothesis of the order in which evolutionary events are assumed to have occurred. Cladistics is today the method of choice to infer phylogenetic trees. The most commonly used methods to infer phylogenies include parsimony, maximum likelihood, and MCMC-based Bayesian inference. Phenetics, popular in the mid-20th century but now largely obsolete, uses distance matrix-based methods to construct trees based on overall similarity which is often assumed to approximate phylogenetic relationships. All methods depend upon an implicit or explicit mathematical model describing the evolution of characters observed in the species included, and are usually used for molecular phylogeny where the characters are aligned nucleotide or amino acid sequences.

–Author

CONTENTS

Chapter–1

INTRODUCTION

In biology, phylogenetics is the study of evolutionary relatedness among various groups of organisms (e.g., species, populations), which is discovered through molecular sequencing data. The term phylogenetics is of Greek origin from the terms *phyle*/ *phylon*, meaning "tribe, race," and genetikos meaning "relative to birth" from genesis "birth". Phylogenetics is that branch of life science which deals with the study of evolutionary relations among various species or populations of organisms, through molecular sequencing data. Taxonomy, the classification of organisms according to similarity, has been richly informed by phylogenetics but remains methodologically and logically distinct The fields overlap however in the science of phylogenetic systematics or cladism, where only phylogenetic trees are used to delimit taxa, each representing a group of lineage-connected individuals

Evolution is regarded as a branching process, whereby populations are altered over time and may speciate into separate branches, hybridize together, or terminate by extinction. This may be visualized as a multi-dimensional character-space that a population moves through over time. The problem posed by phylogenetics is that genetic data are only available for the present, and fossil records (osteometric data) are sporadic and less reliable. Our knowledge of how evolution operates is used to reconstruct the full tree.

There are some terms that describe the nature of a grouping in such trees. For instance, all birds and reptiles are believed to have descended from a single common ancestor, so this taxonomic grouping (yellow in the diagram below) is called monophyletic. "Modern reptile" is a grouping that contains a common ancestor, but does not contain all descendents of that ancestor (birds are excluded). This is an example of a paraphyletic group. A grouping such as warm-blooded animals would include only mammals and birds and is called polyphyletic because the members of this grouping do not include the most recent common ancestor. Thus, a phylogenetic tree is based on a hypothesis of the order in which evolutionary events are assumed to have occurred. Cladistics is today the method of choice to infer phylogenetic trees. The most commonly used methods to infer phylogenies include parsimony, maximum likelihood, and MCMC-based Bayesian inference. Phenetics, popular in the mid-20th century but now largely obsolete, uses distance matrix-based methods to construct trees based on overall similarity which is often assumed to approximate phylogenetic relationships. All methods depend upon an implicit or explicit mathematical model describing the evolution of characters observed in the species included, and are usually used for molecular phylogeny where the characters are aligned nucleotide or amino acid sequences.

The evolutionary connections between organisms are represented graphically through phylogenetic trees. Due to the fact that evolution takes place over long periods of time which can not be observed directly, biologists must continually reconstruct phylogenies by inferring the evolutionary relationships among present-day organisms. Fossils can aid with the reconstruction of phylogenies, however, fossil records are often too poor to be of good help. Therefore, biologists tend to be restricted with analysing present-day organisms to identify their evolutionary relationships. Phylogenetic relationships in the past were reconstructed by looking at phenotypes, often anatomical characteristics. Today, molecular data, which includes protein and DNA sequences, are used to construct phylogenetic trees.

Ernst Haeckel's Recapitulation Theory

During the late 19th century, Ernst Haeckel's recapitulation theory, or biogenetic law, was widely accepted. This theory was often expressed as "ontogeny recapitulates phylogeny", i.e. the development of an organism exactly mirrors the evolutionary development of the species. Haeckel's early version of this hypothesis [that the embryo mirrors adult evolutionary ancestors has since been rejected, and the hypothesis amended as the embryo's development mirroring embryos of its evolutionary ancestors. Most modern biologists recognize numerous connections between ontogeny and phylogeny, explain them using evolutionary theory, or view them as supporting evidence for that theory. Donald Williamson suggested that larvae and embryos represented adults in other taxa that have been transferred by hybridization (the larval transfer theory).

Gene Transfer

Organisms can generally inherit genes in two ways: vertical gene transfer and horizontal gene transfer. Vertical gene transfer is the passage of genes from parent to offspring, and horizontal gene transfer or lateral gene transfer occurs when genes jump between unrelated organisms, a common phenomenon in prokaryotes. Lateral gene transfer has complicated the determination of phylogenies of organisms since inconsistencies have been reported depending on the gene chosen.

Carl Woese came up with the three-domain theory of life (eubacteria, archaea and eukaryotes) based on his discovery that the genes encoding ribosomal RNA are ancient and distributed over all lineages of life with little or no lateral gene transfer. Therefore rRNA are commonly recommended as molecular clocks for reconstructing phylogenies. This has been particularly useful for the phylogeny of microorganisms, to which the species concept does not apply and which are too morphologically simple to be classified based on phenotypic traits.

Taxon Sampling and Phylogenetic Signal

Owing to the development of advanced sequencing techniques in molecular biology, it has become feasible to gather large amounts of data (DNA or amino acid sequences) to estimate phylogenies. For example, it is not rare to find studies with character matrices based on whole mitochondrial genomes. However, it has been proposed that it is more important to increase the number of taxa in the matrix than to increase the number of characters, because the more taxa, the more robust is the resulting phylogeny. This is partly due to the breaking up of long branches. It has been argued that this is an important reason to incorporate data from fossils into phylogenies where possible. Using simulations, Derrick Zwickl and Hillis found that increasing taxon sampling in phylogenetic inference has a positive effect on the accuracy of phylogenetic analyses.

Another important factor that affects the accuracy of tree reconstruction is whether the data analyzed actually contains a useful phylogenetic signal, a term that is used generally to denote whether related organisms tend to resemble each other with respect to their genetic material or phenotypic traits.

Recapitulation Theory

The theory of recapitulation, also called the biogenetic law or embryological parallelism, and often expressed as ontogeny recapitulates phylogeny, was first put forward in 1866 by German zoologist Ernst Haeckel. Haeckel proposed that the embryonal development of an individual organism (its ontogeny) followed the same path as the evolutionary history of its species (its phylogeny). This theory, in the highly elaborate and deterministic form advanced by Haeckel, has, since the early twentieth century, been refuted on many fronts. Haeckel's drawings used artistic licence, his theory was associated with Lamarckism, it was quite clearly wrong in supposing that embryos passed through the adult stages of more primitive life-forms, it ignored organs such as teeth which are "held over" to a late developmental stage, and it was used by Haeckel to promote the supremacy of the white European male. However, the basic idea of recapitulation is still wide spread - Stephen Jay Gould's first book begins by declaring that many medical professionals still believe, privately and informally, that there is "something in" the notion.

Haeckel's Theory

Ontogeny is the growth (size change) and development (shape change) of an individual organism; phylogeny is the evolutionary history of a species. Haeckel's recapitulation theory claims that the development of advanced species was seen to pass through stages represented by adult organisms of more primitive species Otherwise put, each successive stage in the development of an individual represents one of the adult forms that appeared in its evolutionary history.

Haeckel formulated his theory as such: "Ontogeny recapitulates phylogeny". This notion later became simply known as recapitulation (OED: 'a summing up or brief repetition').

Haeckel produced several embryo drawings that often overemphasized similarities between embryos of related species. These found their ways into many biology textbooks, and into popular knowledge.

For example, Haeckel believed that the human embryo with gill slits (pharyngeal arches) in the neck not only signified a fishlike ancestor, but represented an adult "fishlike" developmental stage. Embryonic pharyngeal arches are not gills and do not carry out the same function. They are the invaginations between the gill pouches or pharyngeal pouches, and they open the pharynx to the outside. Gill pouches appear in all tetrapod animal embryos. In mammals, the first gill bar (in the first gill pouch) develops into the lower jaw (Meckel's cartilage), the malleus and the stapes. In a later stage, all gill slits close, with only the ear opening remaining open.

Rejection

Modern biology rejects the literal and universal form of Haeckel's theory. Although humans are generally understood to share ancestors with other taxa, stages of human embryonic development are not functionally equivalent to the adults of these shared common ancestors. In other words, no cleanly defined and functional "fish", "reptile" and "mammal" stages of human embryonal development can be discerned. Moreover, development is non-linear. For example, during kidney development, at one given time, the anterior region of the kidney is less developed (nephridium) than the posterior region (nephron).

Modern biology does recognize numerous connections between ontogeny and phylogeny, and explains them using evolutionary theory without recourse to Haeckel's specific views, and considers them as supporting evidence for that theory.

Historical Influence

Although Haeckel's specific form of recapitulation theory is now discredited among biologists, it had a strong influence on social and educational theories of the late 19th century.

English philosopher Herbert Spencer was one of the most energetic promoters of evolutionary ideas to explain many phenomena. He compactly expressed the basis for a cultural recapitulation theory of education in the following claim.

The maturationist theory of G. Stanley Hall was based on the premise that growing children would recapitulate evolutionary stages of development as they grew up and that there was a one-to-one correspondence between childhood stages and evolutionary history, and that it was counter-productive to push a child ahead of its development stage. The whole notion fit nicely with other social Darwinist concepts, such as the idea that "primitive" societies needed guidance by more advanced societies, i.e. Europe and North America, which were considered by social Darwinists as the pinnacle of evolution. An early form of the law was devised by the 19th-century Estonian zoologist Karl Ernst von Baer, who observed that embryos resemble the embryos, but not the adults, of other species.

Modern Observations

Generally, if a structure pre-dates another structure in evolutionary terms, then it also appears earlier than the other in the embryo. Species which have an evolutionary relationship typically share the early stages of embryonal development and differ in later stages. Examples include:

- The backbone, the common structure among all vertebrates such as fish, reptiles and mammals, appears as one of the earliest structures laid out in all vertebrate embryos.
- The cerebrum in humans, the most sophisticated part of the brain, develops last.

If a structure vanished in an evolutionary sequence, then one can often observe a corresponding structure appearing at one stage during embryonic development, only to disappear or become modified in a later stage. Examples include:

- Whales, which have evolved from land mammals, don't have legs, but tiny remnant leg bones lie buried deep in their bodies. During embryonal development, leg extremities first occur, then recede. Similarly, whale embryos have hair at one stage (like all mammalian embryos), but lose most of it later.
- The common ancestor of humans and monkeys had a tail, and human embryos also have a tail at one point; it later recedes to form the coccyx.
- The swim bladder in fish presumably evolved from a sac connected to the gut, allowing the fish to gulp air. In most modern fish, this connection to the gut has disappeared. In the embryonal development of these fish, the swim bladder originates as an outpocketing of the gut, and is later disconnected from the gut.

Modern Theory

One can explain connections between phylogeny and ontogeny if one assumes that one species changes into another by a sequence of small modifications to its developmental programme (specified by the genome). Modifications that affect early steps of this program will usually require the right modifications in later steps in order to produce an individual that survives and reproduces. Therefore, such successful combinations of changes are less likely to occur and most of the successful changes affect the latest stages of the programme, and the earlier steps are retained. But occasionally, a modification of an earlier step in the programme does succeed: for this reason ontogeny and phylogeny do not strictly correspond, contrary to Haeckel's original theory.

Evolutionary Developmental Biology

Evolutionary developmental biology (evolution of development or informally, evo-devo) is a field of biology that compares the developmental processes of different animals and plants in an attempt to determine the ancestral relationship between organisms and how developmental processes evolved. It addresses the origin and evolution of embryonic development; how modifications of development and developmental processes lead to the production of novel features; the role of developmental plasticity in evolution; how ecology impacts in development and evolutionary change; and the developmental basis of homoplasy and homology.

Although interest in the relationship between ontogeny and phylogeny extends back to the nineteenth century, the contemporary field of evo-devo has gained impetus from the discovery of genes regulating embryonic development in model organisms. General hypotheses remain hard to test because organisms differ so much in shape and form. Nevertheless, it now appears that just as evolution tends to create new genes from parts of old genes (molecular economy), evo-devo demonstrates that evolution alters developmental processes (genes and gene networks) to create new and novel structures from the old gene networks (such as bone structures of the jaw deviating to the ossicles of the middle ear) or will conserve (molecular economy) a similar program in a host of organisms such as eye development genes in molluscs, insects, and vertebrates Initially the major interest has been in the macroevolutionary evidence of homology in the cellular and molecular mechanisms that regulate body plan and organ development. However, more modern approaches include microevolution and developmental changes associated in speciation.

Basic Principles

Charles Darwin's theory of evolution is based on three principles: natural selection, heredity, and variation. At the time that Darwin wrote, the principles underlying heredity and variation were poorly understood. In the 1940s, however, biologists incorporated Gregor Mendel's principles of genetics to explain both, resulting in the modern synthesis. It was not until the 1980s and 1990s, however, when more comparative molecular sequence data between different kinds of organisms was amassed and detailed, that an understanding of the molecular basis of the developmental mechanisms has arisen. Currently, it is well understood how genetic mutation occurs. However, developmental mechanisms are not understood sufficiently to explain which kinds of phenotypic variation can arise in each generation from variation at the genetic level. Evolutionary developmental biology studies how the dynamics of development determine the phenotypic variation arising from genetic variation and how that affects phenotypic evolution (specially its direction). At the same time evolutionary developmental biology also studies how development itself evolves. Thus, the origins of evolutionary developmental biology come from both an improvement in molecular biology techniques as applied to development, and the full appreciation of the limitations of classic neo-Darwinism as applied to phenotypic evolution. Some evo-devo researchers see themselves as extending and enhancing the modern synthesis by incorporating into it findings of molecular genetics and developmental biology. Others, drawing on findings of discordances between genotype and phenotype and epigenetic mechanisms of development, are mounting an explicit challenge to neo-Darwinism.

Evolutionary developmental biology is not yet a unified discipline, but can be distinguished from earlier approaches to evolutionary theory by its focus on a few crucial ideas. One of these is modularity: as has been long recognized, plants and animal bodies are modular: they are organized into developmentally and anatomically distinct parts.

Often these parts are repeated, such as fingers, ribs, and body segments. Evo-devo seeks the genetic and evolutionary basis for the division of the embryo into distinct modules, and for the partly independent development of such modules. Another central idea is that some gene products function as switches whereas others act as diffusible signals. Genes specify proteins, some of which act as structural components of cells and others as enzymes that regulate various biochemical pathways within an organism. Most biologists working within the modern synthesis assumed that an organism is a straightforward reflection of its component genes. The modification of existing, or evolution of new, biochemical pathways (and, ultimately, the evolution of new species of organisms) depended on specific genetic mutations.

In 1961, however, Jacques Monod, Jean-Pierre Changeux and François Jacob discovered within the bacterium Escherichia coli a gene that functioned only when "switched on" by an environmental stimulus. Later, scientists discovered specific genes in animals, including a subgroup of the genes which contain the homeobox DNA motif, called Hox genes, that act as switches for other genes, and could be induced by other gene products, morphogens, that act analogously to the external stimuli in bacteria.

These discoveries drew biologists' attention to the fact that genes can be selectively turned on and off, rather than being always active, and that highly disparate organisms (for example, fruit flies and human beings) may use the same genes for embryogenesis (e.g., the genes of the "developmental-genetic toolkit"), just regulating them differently. Similarly, organismal form can be influenced by mutations in promoter regions of genes, those DNA sequences at which the products of some genes bind to and control the activity of the same or other genes, not only protein-specifying sequences.

In addition to providing new support for Darwin's assertion that all organisms are descended from a common ancestor, this finding suggested that the crucial distinction between different species (even different orders or phyla) may be due less to differences in their content of gene products than to differences in spatial and temporal expression of conserved genes. The implication that large evolutionary changes in body morphology are associated with changes in gene regulation, rather than the evolution of new genes, suggested that the action of natural selection on promoters responsive to Hox and other "switch" genes may play a major role in evolution. Another focus of evo-devo is developmental plasticity, the basis of the recognition that organismal phenotypes are not uniquely determined by their genotypes.

If generation of phenotypes is conditional, and dependent on external or environmental inputs, evolution can proceed by a "phenotype-first" route, with genetic change following, rather than initiating, the formation of morphological and other phenotypic novelties. In the Origin of Species (1859), Charles Darwin proposed evolution through natural selection, a theory central to modern biology. Darwin recognised the importance of embryonic development in the understanding of evolution:

> "We can see why characters derived from the embryo should be of equal importance with those derived from the adult, for a natural classification of course includes all ages."

Ernst Haeckel (1866), in response to Darwin's theory, proposed that "ontogeny recapitulates phylogeny", that is, the development of the embryo of every species (ontogeny) fully repeats the evolutionary development of that species (phylogeny). Haeckel's concept explained, for example, why humans, and indeed all vertebrates, have gill slits and tails early in embryonic development. While true in general, his theory has been discredited in its absolute form. However, it served as a backdrop for a renewed interest in the evolution of development after the modern evolutionary synthesis was established (roughly 1936 to 1947).

Stephen Jay Gould called this approach to explaining evolution as terminal addition; as if every evolutionary advance was added as new stage by reducing the duration of the older stages. The idea was based on observations of neoteny his was extended by the more general idea of heterochrony (changes in timing of development) as a mechanism for evolutionary change.

D'Arcy Thompson postulated that differential growth rates could produce variations in form in his 1917 book *On Growth and Form*. He showed the underlying similarities in body plans and how geometric transformations could be used to explain the variations.

Edward B. Lewis discovered homeotic genes, rooting the emerging discipline of evo-devo in molecular genetics. In 2000, a special section of the Proceedings of the National Academy of Sciences (PNAS) was devoted to "evo-devo", and an entire 2005 issue of the Journal of Experimental Zoology Part B: Molecular and Developmental Evolution was devoted to the key evo-devo topics of evolutionary innovation and morphological novelty.

The Developmental-genetic Toolkit

The developmental-genetic toolkit consists of genes whose products control the development of a multicellular organism. Differences in deployment of toolkit genes affect the body plan and the number, identity, and pattern of body parts. The toolkit is highly conserved across animal phyla. Only a small fraction of the genes in the genome are involved in development. The majority of toolkit genes are components of signaling pathways, and encode for the production of transcription factors, cell adhesion proteins, cell surface receptor proteins, and secreted morphogens. Their function is highly correlated with their spatial and temporal expression patterns. One of the major goals of evo-devo is to catalogue all genes (their identity, product, function, and interaction) in the toolkit.

Among the most important of the toolkit genes are those of the Hox gene cluster, or complex. Hox genes, transcription factors containing the more broadly distributed homeobox protein-binding DNA motif, function in patterning the body axis. Thus, by combinatorially specifying the identity of particular body regions, Hox genes determine where limbs and other body segments will grow in a developing embryo or larva. Mutations in any one of these genes can lead to the growth of extra, typically non-functional body parts in invertebrates, for example aristapedia complex in Drosophila, which results in a leg growing from the head in place of an antenna and is due to a defect in a single gene (this mutation is also known as Antennapedia).

Development and the Origin of Novelty

Among the more surprising and, perhaps, counter-intuitive (from a neo-Darwinian viewpoint) results of recent research in evolutionary developmental biology is that the diversity of body plans and morphology in organisms across many phyla are not necessarily reflected in diversity at the level of the sequences of genes, including those of the developmental genetic toolkit and other genes involved in development. Indeed, as Gerhart and Kirschner have noted, there is an apparent paradox: "where we most expect to find variation, we find conservation, a lack of change".

Even within a species, the occurrence of novel forms within a population does not generally correlate with levels of genetic variation sufficient to account for all morphological diversity. For example, there is significant variation in limb morphologies amongst salamanders and in differences in segment number in centipedes, even when the respective genetic variation is low.

A major question then, for evo-devo studies, is: If the morphologicl novelty we observe at the level of different clades is not always reflected in the genome, where does it come from? Apart from neo-Darwinian mechanisms such as mutation, translocation and duplication of genes, novelty may also arise by mutation-driven changes in gene regulation. The finding that much biodiversity is not due to differences in genes, but rather to alterations in gene regulation, has introduced an important new element into evolutionary theory Diverse organisms may have highly conserved developmental genes, but highly divergent regulatory mechanisms for these genes. Changes in gene regulation are "second-order" effects of genes, resulting from the interaction and timing of activity of gene networks, as distinct from the functioning of the individual genes in the network.

The discovery of the homeotic Hox gene family in vertebrates in the 1980s allowed researchers in developmental biology to empirically assess the relative roles of gene duplication and gene regulation with respect to their importance in the evolution of morphological diversity. Several biologists, including Sean B. Carroll of the University of Wisconsin-Madison suggest that "changes in the cis-regulatory systems of genes" are more significant than "changes in gene number or protein function". These researchers argue that the combinatorial nature of transcriptional regulation allows a rich substrate for morphological diversity, since variations in the level, pattern, or timing of gene expression may provide more variation for natural selection to act upon than changes in the gene product alone.

Epigenetic alterations of gene regulation or phenotype generation that are subsequently consolidated by changes at the gene level constitute another class of mechanisms for evolutionary innovation. Epigenetic changes include modification of the genetic material due to methylation and other reversible chemical alteration as well as nonprogrammed remolding of the organism by physical and other environmental effects due to the inherent plasticity of developmental mechanisms The biologists Stuart A. Newman and Gerd B. Müller have suggested that organisms early in the history of multicellular life were more susceptible to this second category of epigenetic determination than are modern organisms, providing a basis for early macroevolutionary changes.

Chapter–2

GENETICS

INTRODUCTION

Genetics studies how living organisms inherit features from their ancestors – for example, children often look like their parents. Genetics tries to identify which features are inherited, and work out the details of how these features are passed from generation to generation.

In genetics, a feature of an organism is called a "trait". Some traits are features of an organism's physical appearance, for example, a person's eye-color, height or weight. There are many other types of traits and these range from aspects of behavior to resistance to disease. Traits are often inherited, for example tall and thin people tend to have tall and thin children. Other traits come from the interaction between inherited features and the environment. For example a child might inherit the tendency to be tall, but if there is very little food where they live and they are poorly nourished, they will still be short. The way genetics and environment interact to produce a trait can be complicated: for example, the chances of somebody dying of cancer or heart disease seem to depend on both their family history and their lifestyle.

Genetic information is carried by a long molecule called DNA which is copied and inherited across generations. Traits are carried in DNA as instructions for constructing and operating an organism. These instructions are contained in segments of DNA called genes. DNA is made of a sequence of simple units, with the order of these units spelling out instructions in the genetic code. This is similar to the orders of letters spelling out words.

The organism "reads" the sequence of these units and decodes the instruction. Not all the genes for a particular instruction are exactly the same. Different forms of one type of gene are called different alleles of that gene. As an example, one allele of a gene for hair color could carry the instruction to produce a lot of the pigment in black hair, while a different allele could give a garbled version of this instruction, so that no pigment is produced and the hair is white.

Mutations are random events that change the sequence of a gene and therefore create a new allele. Mutations can produce a new trait, such as turning an allele for black hair into an allele for white hair. The appearance of new traits is important in evolution.

GENES AND INHERITANCE

Genes are inherited as units, with parents dividing out their genes to their offspring. You can think of this process like mixing two hands of cards, shuffling them, and then dealing them out again. Humans have two copies of each of their genes and when people reproduce they make copies of their genes in eggs or sperm, but only put in one copy of each type of gene.

An egg then joins with a sperm to give a child with a new set of genes. This child will have the same number of genes as its parents but for any gene one of their two copies will come from the father, and one from the mother. The effects of this mixing depends on the types (the alleles) of the gene you are interested in.

If the father has two alleles specifying green eyes, and the mother has two alleles specifying brown eyes, all their children will get two alleles giving different instructions, one for green eyes and one for brown. The eye color of these children depends on how these alleles work together. If one allele overrides the instructions from another, it is called the dominant allele, and the allele that is overridden is called the recessive allele. In the case of a daughter with both green and brown alleles, brown is dominant and she ends up with brown eyes.

Green eyes are a recessive trait. However, the green eye color allele is still there in this brown-eyed girl, it just doesn't show. This is a difference between what you see on the surface (the set of observable traits of an organism, also called its phenotype) and which genes are in this organism (its genotype). In this example you can call the brown allele "B" and the green allele "g". (It is normal to write dominant alleles with capital letters and recessive ones with lower-case letters.) The brown-eyed daughter has the "brown eye phenotype" but her genotype is Bg, with one copy of the B allele, and one of the g allele.

Now imagine that this woman grows up and has children with a brown-eyed man who also has a Bg genotype. Her eggs will be a mixture of two types, one sort containing the B allele, and one sort the gallele. Similarly, her partner will produce a mix of two types of sperm containing one or the other of the two alleles. Now, when the alleles are mixed up in the offspring, these children have a chance of getting either brown or green eyes, since they could get a genotype of BB = brown eyes, Bg = brown eyes or gg = green eyes. In this generation, there is therefore a chance of the recessive allele showing itself in the phenotype of the children - some of them may have green eyes like their grandfather.

Many traits are inherited in a more complicated way than the example above. This can happen when there are several genes involved, each contributing a small part to the end result. Tall people tend to have tall children because their children get a package of many alleles that each contribute a bit to how much they grow. However, there are not clear groups of "short people" and "tall people", like there are groups of people with brown or green eyes. This is because of the large number of genes involved; this makes the trait very variable and people are many different heights Inheritance can also be complicated when the trait depends on the interaction between genetics and the environment. This is quite common, for example, if a child does not eat enough nutritious food this will not change traits like eye color, but it could stunt their growth.

Inherited Diseases

Some diseases are hereditary and run in families; others, such as infectious diseases, are caused by the environment. Other disorders are caused by a combination of hereditary and environmental factors.

Diseases that are caused by a single allele of a gene and are inherited in families are called genetic disorders. These include diseases like Huntington's disease, Cystic fibrosis or Duchenne muscular dystrophy. Cystic fibrosis, for example, is caused by mutations in a single gene called CFTR and is inherited as a recessive trait Other diseases are influenced by genetics, but which alleles a person gets from their parents only changes their risk of getting a disease. Most of these diseases are inherited in a complex way, with either multiple genes involved, or both genes and the environment being important.

As an example, the risk of breast cancer is 50 times higher in the families most at risk, compared to the families least at risk. This variation is probably due to a large number of alleles. Each of them changes the risk a little bit. Several of the genes involved have been identified, such as BRCA1 and BRCA2, but not all of them. However, although some of the risk is genetic, being overweight, drinking a lot of alcohol, or not exercising, all increase the risk of this cancer A woman's risk of breast cancer is therefore the result of a large number of alleles and her environment, so it is very hard to predict.

GENES MAKE PROTEINS

The function of genes is to provide the information needed to make molecules called proteins in cells Cells are the smallest independent parts of organisms: the human body contains about 100 trillion cells, while very small organisms like bacteria are just a single cell. A cell is like a miniature and very complex factory that can make all the parts needed to produce a copy of itself, which happens when cells divide. There is a simple division of labor in cells - genes give instructions and proteins carry out these instructions, tasks like building a new copy of a cell, or repairing damage Each type of protein is a specialist that only does one job, so if a cell needs to do something new, it must make a new protein to do this job. Similarly, if a cell needs to do something faster or slower than before, it makes more or less of the protein responsible. Genes tell cells what to do by telling them which proteins to make and in what amounts.

Proteins are made of a chain of 20 different types of amino acids. This chain folds up into a compact shape, rather like an untidy ball of rope. The shape of the protein is determined by the sequence of amino acids along its chain and it is this shape that, in turn, determines what the protein will do For example, some proteins have depressions in their surface that perfectly match another molecule, allowing the protein to bind to this molecule very tightly. Other proteins are enzymes, which are like tiny machines that can alter other molecules.

The information in DNA is held in the sequence of the repeating units along the DNA chain These units are four types of nucleotides (A,T,G and C) and the sequence of nucleotides stores information in an alphabet called the genetic code. When a gene is read by a cell the DNA sequence is copied into a very similar molecule called RNA (this process is called transcription). Transcription is controlled by other DNA sequences (such as promoters), which show a cell where genes are, and control how often they are copied. The RNA copy made from a gene is then fed through a structure called a ribosome, which translates the sequence of nucleotides in the RNA into the correct sequence of amino acids and joins these amino acids together to make a complete protein chain. The new protein then folds up into its active form. The process of moving information from the language of DNA into the language of amino acids is called translation.

DNA is unwound and nucleotides are matched to make two new strands.If the sequence of the nucleotides in a gene changes, the sequence of the amino acids in the protein it produces may also change - if part of a gene is deleted, the protein produced will be shorter and may not work any more. This is the reason why different alleles of a gene can have different effects in an organism. As an example, hair color depends on how much of a dark substance called melanin is put into the hair as it grows. If a person has a normal set of the genes involved in making melanin, they make all the proteins needed and they grow dark hair. However, if the alleles for a particular protein have different sequences and produce proteins that do not do the job correctly, no melanin will be produced and the hair will be white. This condition is called albinism and the person with this condition is called an albino.

Copies of Genes are Inherited

When genes are passed from a parent to a child they are copied - the parent keeps the same number of genes as they had before and just passes on the new copies to their offspring. Genes are also copied each time a cell divides into two new cells. The process that copies DNA is called DNA replication.

The DNA can be copied very easily and accurately because each piece of DNA can direct the creation of a new copy of its information. This is because DNA is made of two strands that pair together like the two sides of a zipper. The nucleotides are in the center, like the teeth in the zipper, and pair up to hold the two strands together. Importantly, the four different sorts of nucleotides are different shapes, so in order for the strands to close up properly, an A nucleotide must go opposite a T nucleotide, and a G opposite a C. This exact pairing is called base pairing.

When DNA is copied, the two strands of the old DNA are pulled apart by enzymes which move along each of the two single strands pairing up new nucleotide units and then zipping the strands closed. This produces two new pieces of DNA, each containing one strand from the old DNA and one newly made strand. This process isn't perfect and sometimes the proteins will make mistakes and put the wrong nucleotide into the strand they are building. This causes a change in the sequence of that gene. These changes in DNA sequence are called mutations Mutations produce new alleles of genes. Sometimes these changes stop the gene from working properly, like the melanin genes discussed above. In other cases these mutations can change what the gene does or even let it do its job a little better than before. These mutations and their effects on the traits of organisms are one of the causes of evolution.

GENES AND EVOLUTION

A population of organisms evolves when an inherited trait becomes more common or less common over time. For instance, all the mice living on an island would be a single population of mice. If over a few generations, white mice went from being rare, to being a large part of this population, then the coat color of these mice would be evolving. In terms of genetics, this is called a change in allele frequency—such as an increase in the frequency of the allele for white fur.

Alleles become more or less common either just by chance (in a process called genetic drift), or through natural selection. In natural selection, if an allele makes it more likely that an organism will survive and reproduce, then over time this allele will become more common. But if an allele is harmful, natural selection will make it less common. For example, if the island was getting colder each year and was covered with snow for much of the time, then the allele for white fur would become useful for the mice, since it would make them harder to see against the snow. Fewer of the white mice would be eaten by predators, so over time white mice would out-compete mice with dark fur. White fur alleles would become more common, and dark fur alleles would become more rare.

Mutations Create New Alleles

These alleles have new DNA sequences and can produce proteins with new properties. So if an island was populated entirely by black mice, mutations could happen creating alleles for white fur. The combination of mutations creating new alleles at random, and natural selection picking out those which are useful, causes adaptation. This is when organisms change in ways that help them to survive and reproduce.

Genetic Engineering

Since traits come from the genes in a cell, if you put a new piece of DNA into a cell, this can produce a new trait. This is how genetic engineering works. For example, crop plants can be given a gene from an Arctic fish, so they produce an antifreeze protein in their leaves. This can help prevent frost damage. Other genes that can be put into crops include a natural insecticide from the bacteria *Bacillus thuringiensis* (BT). The insecticide kills insects that eat the plants, but is harmless to people. In these plants the new genes are put into the plant before it is grown, so the genes will be in every part of the plant, including its seeds. The plant's offspring will then inherit the new genes, something which has led to concern about the spread of new traits into wild plants.

The kind of technology used in genetic engineering is also being developed to treat people with genetic disorders in an experimental medical technique called gene therapy However, here the new gene is put in after the person has grown up and become ill, so any new gene will not be inherited by their children. Gene therapy works by trying to replace the allele that causes the disease with an allele that will work properly.

Genetics, a discipline of biology, is the science of heredity and variation in living organisms. The fact that living things inherit traits from their parents has been used since prehistoric times to improve crop plants and animals through selective breeding. However, the modern science of genetics, which seeks to understand the process of inheritance, only began with the work of Gregor Mendel in the mid-nineteenth century. Although he did not know the physical basis for heredity, Mendel observed that organisms inherit traits in a discrete manner—these basic units of inheritance are now called genes.

The DNA, the molecular basis for inheritance. Each strand of DNA is a chain of nucleotides, matching each other in the center to form what look like rungs on a twisted ladder.Genes correspond to regions within DNA, a molecule composed of a chain of four different types of nucleotides—the sequence of these nucleotides is the genetic information organisms inherit. DNA naturally occurs in a double stranded form, with nucleotides on each strand complementary to each other. Each strand can act as a template for creating a new partner strand—this is the physical method for making copies of genes that can be inherited.

The sequence of nucleotides in a gene is translated by cells to produce a chain of amino acids, creating proteins—the order of amino acids in a protein corresponds to the order of nucleotides in the gene. This is known as the genetic code. The amino acids in a protein determine how it folds into a three-dimensional shape; this structure is, in turn, responsible for the protein's function. Proteins carry out almost all the functions needed for cells to live. A change to the DNA in a gene can change a protein's amino acids, changing its shape and function: this can have a dramatic effect in the cell and on the organism as a whole.

Although genetics plays a large role in the appearance and behavior of organisms, it is the combination of genetics with what an organism experiences that determines the ultimate outcome. For example, while genes play a role in determining a person's height, the nutrition and health that person experiences in childhood also have a large effect.

Although the science of genetics began with the applied and theoretical work of Gregor Mendel in the mid-1800s, other theories of inheritance preceded Mendel. A popular theory during Mendel's time was the concept of blending inheritance: the idea that individuals inherit a smooth blend of traits from their parents. Mendel's work disproved this, showing that traits are composed of combinations of distinct genes rather than a continuous blend. Another theory that had some support at that time was the inheritance of acquired characteristics: the belief that individuals inherit traits strengthened by their parents. This theory (commonly associated with Jean-Baptiste Lamarck) is now known to be wrong—the experiences of individuals do not affect the genes they pass to their children Other theories included the pangenesis of Charles Darwin (which had both acquired and inherited aspects) and Francis Galton's reformulation of pangenesis as both particulate and inherited.

MENDELIAN AND CLASSICAL GENETICS

The modern science of genetics traces its roots to Gregor Johann Mendel, a German-Czech Augustinian monk and scientist who studied the nature of inheritance in plants. In his paper "Versuche über Pflanzenhybriden" ("Experiments on Plant Hybridization"), presented in 1865 to the Naturforschender Verein (Society for Research in Nature) in Brünn, Mendel traced the inheritance patterns of certain traits in pea plants and described them mathematically Although this pattern of inheritance could only be observed for a few traits, Mendel's work suggested that heredity was particulate, not acquired, and that the inheritance patterns of many traits could be explained through simple rules and ratios.

The importance of Mendel's work did not gain wide understanding until the 1890s, after his death, when other scientists working on similar problems re-discovered his research. William Bateson, a proponent of Mendel's work, coined the word genetics in 1905. (The adjective genetic, derived from the Greek word genesis - , "origin" and that from the word genno - "to give birth", predates the noun and was first used in a biological sense in 1860 Bateson popularized the usage of the word genetics to describe the study of inheritance in his inaugural address to the Third International Conference on Plant Hybridization in London, England, in 1906.

After the rediscovery of Mendel's work, scientists tried to determine which molecules in the cell were responsible for inheritance. In 1910, Thomas Hunt Morgan argued that genes are on chromosomes, based on observations of a sex-linked white eye mutation in fruit flies. In 1913, his student Alfred Sturtevant used the phenomenon of genetic linkage to show that genes are arranged linearly on the chromosome.

MOLECULAR GENETICS

James D. Watson (pictured) and Francis Crick determined the structure of DNA in 1953.Although genes were known to exist on chromosomes, chromosomes are composed of both protein and DNA—scientists did not know which of these was responsible for inheritance. In 1928, Frederick Griffith discovered the phenomenon of transformation: dead bacteria could transfer genetic material to "transform" other still-living bacteria. Sixteen years later, in 1944, Oswald Theodore Avery, Colin McLeod and Maclyn McCarty identified the molecule responsible for transformation as DNA. The Hershey-Chase experiment in 1952 also showed that DNA (rather than protein) was the genetic material of the viruses that infect bacteria, providing further evidence that DNA was the molecule responsible for inheritance.

James D. Watson and Francis Crick determined the structure of DNA in 1953, using the X-ray crystallography work of Rosalind Franklin that indicated DNA had a helical structure (i.e., shaped like a corkscrew). Their double-helix model had two strands of DNA with the nucleotides pointing inward, each matching a complementary nucleotide on the other strand to form what looks like rungs on a twisted ladder. This structure showed that genetic information exists in the sequence of nucleotides on each strand of DNA. The structure also suggested a simple method for duplication: if the strands are separated, new partner strands can be reconstructed for each based on the sequence of the old strand.

Although the structure of DNA showed how inheritance worked, it was still not known how DNA influenced the behavior of cells. In the following years, scientists tried to understand how DNA controls the process of protein production. It was discovered that the cell uses DNA as a template to create matching messenger RNA (a molecule with nucleotides, very similar to DNA). The nucleotide sequence of a messenger RNA is used to create an amino acid sequence in protein; this translation between nucleotide and amino acid sequences is known as the genetic code.

With this molecular understanding of inheritance, an explosion of research became possible. One important development was chain-termination DNA sequencing in 1977 by Frederick Sanger: this technology allows scientists to read the nucleotide sequence of a DNA molecule. In 1983, Kary Banks Mullis developed the polymerase chain reaction, providing a quick way to isolate and amplify a specific section of a DNA from a mixture. Through the pooled efforts of the Human Genome Project and the parallel private effort by Celera Genomics, these and other techniques culminated in the sequencing of the human genome in 2003.

FEATURES OF INHERITANCE

At its most fundamental level, inheritance in organisms occurs by means of discrete traits, called genes. This property was first observed by Gregor Mendel, who studied the segregation of heritable traits in pea plants In his experiments studying the trait for flower colour, Mendel observed that the flowers of each pea plant were either purple or white - and never an intermediate between the two colours. These different, discrete versions of the same gene are called alleles.

In the case of pea plants, each organism has two alleles of each gene, and the plants inherit one allele from each parent. Many organisms, including humans, have this pattern of inheritance. Organisms with two copies of the same allele are called homozygous, while organisms with two different alleles are heterozygous.

The set of alleles for a given organism is called its genotype, while the observable trait the organism has is called its phenotype. When organisms are heterozygous, often one allele is called dominant as its qualities dominate the phenotype of the organism, while the other allele is called recessive as its qualities recede and are not observed. Some alleles do not have complete dominance and instead have incomplete dominance by expressing an intermediate phenotype, or codominance by expressing both alleles at once.

When a pair of organisms reproduce sexually, their offspring randomly inherit one of the two alleles from each parent. These observations of discrete inheritance and the segregation of alleles are collectively known as Mendel's first law or the Law of Segregation.

Geneticists use diagrams and symbols to describe inheritance. A gene is represented by a letter (or letters)—the capitalized letter represents the dominant allele and the recessive is represented by lowercase Often a "+" symbol is used to mark the usual, non-mutant allele for a gene.

In fertilization and breeding experiments (and especially when discussing Mendel's laws) the parents are referred to as the "P" generation and the offspring as the "F1" (first filial) generation. When the F1 offspring mate with each other, the offspring are called the "F2" (second filial) generation. One of the common diagrams used to predict the result of cross-breeding is the Punnett square. When studying human genetic diseases, geneticists often use pedigree charts to represent the inheritance of traits. These charts map the inheritance of a trait in a family tree.

INTERACTIONS OF MULTIPLE GENES

Human height is a complex genetic trait. Francis Galton's data from 1889 shows the relationship between offspring height as a function of mean parent height. While correlated, remaining variation in offspring heights indicates environment is also an important factor in this trait. Organisms have thousands of genes, and in sexually reproducing organisms assortment of these genes are generally independent of each other. This means that the inheritance of an allele for yellow or green pea color is unrelated to the inheritance of alleles for white or purple flowers. This phenomenon, known as "Mendel's second law" or the "Law of independent assortment", means that the alleles of different genes get shuffled between parents to form offspring with many different combinations

Often different genes can interact in a way that influences the same trait. In the Blue-eyed Mary (*Omphalodes verna*), for example, there exists a gene with alleles that determine the colour of flowers: blue or magenta. Another gene, however, controls whether the flowers have color at all: color or white. When a plant has two copies of this white allele, its flowers are white - regardless of whether the first gene has blue or magenta alleles. This interaction between genes is called epistasis, with the second gene epistatic to the first.

Many traits are not discrete features (e.g. purple or white flowers) but are instead continuous features (e.g. human height and skin colour). These complex traits are the product of many genes. The influence of these genes is mediated, to varying degrees, by the environment an organism has experienced. The degree to which an organism's genes contribute to a complex trait is called heritability. Measurement of the heritability of a trait is relative - in a more variable environment, the environment has a bigger influence on the total variation of the trait. For example, human height is a complex trait with a heritability of 89% in the United States. In Nigeria, however, where people experience a more variable access to good nutrition and health care, height has a heritability of only 62%.

The molecular structure of DNA. Bases pair through the arrangement of hydrogen bonding between the strands.The molecular basis for genes is deoxyribonucleic acid (DNA). DNA is composed of a chain of nucleotides, of which there are four types: adenine (A), cytosine (C), guanine (G), and thymine (T). Genetic information exists in the sequence of these nucleotides, and genes exist as stretches of sequence along the DNA chain. Viruses are the only exception to this rule—sometimes viruses use the very similar molecule RNA instead of DNA as their genetic material.

The DNA normally exists as a double-stranded molecule, coiled into the shape of a double-helix. Each nucleotide in DNA preferentially pairs with its partner nucleotide on the opposite strand: A pairs with T, and C pairs with G. Thus, in its two-stranded form, each strand effectively contains all necessary information, redundant with its partner strand. This structure of DNA is the physical basis for inheritance: DNA replication duplicates the genetic information by splitting the strands and using each strand as a template for synthesis of a new partner strand.

Genes are arranged linearly along long chains of DNA sequence, called chromosomes. In bacteria, each cell has a single circular chromosome, while eukaryotic organisms (which includes plants and animals) have their DNA arranged in multiple linear chromosomes. These DNA strands are often extremely long; the largest human chromosome, for example, is about 247 million base pairs in length. The DNA of a chromosome is associated with structural proteins that organize, compact, and control access to the DNA, forming a material called chromatin; in eukaryotes, chromatin is usually composed of nucleosomes, repeating units of DNA wound around a core of histone proteins. The full set of hereditary material in an organism (usually the combined DNA sequences of all chromosomes) is called the genome.

While haploid organisms have only one copy of each chromosome, most animals and many plants are diploid, containing two of each chromosome and thus two copies of every gene The two alleles for a gene are located on identical loci of sister chromatids, each allele inherited from a different parent.

Chromosomes are copied, condensed, and organized. Then, as the cell divides, chromosome copies separate into the daughter cells.An exception exists in the sex chromosomes, specialized chromosomes many animals have evolved that play a role in determining the sex of an organism In humans and other mammals, the Y chromosome has very few genes and triggers the development of male sexual characteristics, while the X chromosome is similar to the other chromosomes and contains many genes unrelated to sex determination. Females have two copies of the X chromosome, but males have one Y and only one X chromosome - this difference in X chromosome copy numbers leads to the unusual inheritance patterns of sex-linked disorders.

Reproduction

When cells divide, their full genome is copied and each daughter cell inherits one copy. This process, called mitosis, is the simplest form of reproduction and is the basis for asexual reproduction. Asexual reproduction can also occur in multicellular organisms, producing offspring that inherit their genome from a single parent. Offspring that are genetically identical to their parents are called clones.

Eukaryotic organisms often use sexual reproduction to generate offspring that contain a mixture of genetic material inherited from two different parents. The process of sexual reproduction alternates between forms that contain single copies of the genome (haploid) and double copies (diploid). Haploid cells fuse and combine genetic material to create a diploid cell with paired chromosomes. Diploid organisms form haploids by dividing, without replicating their DNA, to create daughter cells that randomly inherit one of each pair of chromosomes. Most animals and many plants are diploid for most of their lifespan, with the haploid form reduced to single cell gametes.

Although they do not use the haploid/diploid method of sexual reproduction, bacteria have many methods of acquiring new genetic information. Some bacteria can undergo conjugation, transferring a small circular piece of DNA to another bacterium Bacteria can also take up raw DNA fragments found in the environment and integrate them into their genome, a phenomenon known as transformation. These processes result in horizontal gene transfer, transmitting fragments of genetic information between organisms that would be otherwise unrelated.

Recombination and Linkage

Thomas Hunt Morgan's 1916 illustration of a double crossover between chromosomesThe diploid nature of chromosomes allows for genes on different chromosomes to assort independently during sexual reproduction, recombining to form new combinations of genes. Genes on the same chromosome would theoretically never recombine, however, were it not for the process of chromosomal crossover. During crossover, chromosomes exchange stretches of DNA, effectively shuffling the gene alleles between the chromosomes This process of chromosomal crossover generally occurs during meiosis, a series of cell divisions that creates haploid cells.

The probability of chromosomal crossover occurring between two given points on the chromosome is related to the distance between them. For an arbitrarily long distance, the probability of crossover is high enough that the inheritance of the genes is effectively uncorrelated. For genes that are closer together, however, the lower probability of crossover means that the genes demonstrate genetic linkage - alleles for the two genes tend to be inherited together. The amounts of linkage between a series of genes can be combined to form a linear linkage map that roughly describes the arrangement of the genes along the chromosome.

Gene Expression

The genetic code: DNA, through a messenger RNA intermediate, codes for protein with a triplet code.Genes generally express their functional effect through the production of proteins, which are complex molecules responsible for most functions in the cell. Proteins are chains of amino acids, and the DNA sequence of a gene (through RNA intermediate) is used to produce a specific protein sequence. This process begins with the production of an RNA molecule with a sequence matching the gene's DNA sequence, a process called transcription.

This messenger RNA molecule is then used to produce a corresponding amino acid sequence through a process called translation. Each group of three nucleotides in the sequence, called a codon, corresponds to one of the twenty possible amino acids in protein - this correspondence is called the genetic code. The flow of information is unidirectional: information is transferred from nucleotide sequences into the amino acid sequence of proteins, but it never transfers from protein back into the sequence of DNA—a phenomenon Francis Crick called the central dogma of molecular biology.

The dynamic structure of hemoglobin is responsible for its ability to transport oxygen within mammalian blood.

A single amino acid change causes hemoglobin to form fibers.The specific sequence of amino acids results in a unique three-dimensional structure for that protein, and the three-dimensional structures of protein are related to their function Some are simple structural molecules, like the fibres formed by the protein collagen. Proteins can bind to other proteins and simple molecules, sometimes acting as enzymes by facilitating chemical reactions within the bound molecules (without changing the structure of the protein itself). Protein structure is dynamic; the protein hemoglobin bends into slightly different forms as it facilitates the capture, transport, and release of oxygen molecules within mammalian blood.

A single nucleotide difference within DNA can cause a single change in the amino acid sequence of a protein. Because protein structures are the result of their amino acid sequences, some changes can dramatically change the properties of a protein by destabilizing the structure or changing the surface of the protein in a way that changes its interaction with other proteins and molecules. For example, sickle-cell anemia is a human genetic disease that results from a single base difference within the coding region for the ß-globin section of hemoglobin, causing a single amino acid change that changes hemoglobin's physical properties. Sickle-cell versions of hemoglobin stick to themselves, stacking to form fibers that distort the shape of red blood cells carrying the protein. These sickle-shaped cells no longer flow smoothly through blood vessels, having a tendency to clog or degrade, causing the medical problems associated with this disease.

Some genes are transcribed into RNA but are not translated into protein products - these are called non-coding RNA molecules. In some cases, these products fold into structures which are involved in critical cell functions (eg. ribosomal RNA and transfer RNA). RNA can also have regulatory effect through hybridization interactions with other RNA molecules (e.g. microRNA).

Nature Versus Nurture

Siamese cats have a temperature-sensitive mutation in pigment production. Although genes contain all the information an organism uses to function, the environment plays an important role in determining the ultimate phenotype—a dichotomy often referred to as "nature vs. nurture". The phenotype of an organism depends on the interaction of genetics with the environment. One example of this is the case of temperature-sensitive mutations. Often, a single amino acid change within the sequence of a protein does not change its behavior and interactions with other molecules, but it does destabilize the structure. In a high temperature environment, where molecules are moving more quickly and hitting each other, this results in the protein losing its structure and failing to function. In a low temperature environment, however, the protein's structure is stable and functions normally. This type of mutation is visible in the coat coloration of Siamese cats, where a mutation in an enzyme responsible for pigment production causes it to destabilize and lose function at high temperatures. The protein remains functional in areas of skin that are colder—legs, ears, tail, and face—and so the cat has dark fur at its extremities.

Environment also plays a dramatic role in effects of the human genetic disease phenylketonuria. The mutation that causes phenylketonuria disrupts the ability of the body to break down the amino acid phenylalanine, causing a toxic build-up of an intermediate molecule that, in turn, causes severe symptoms of progressive mental retardation and seizures. If someone with the phenylketonuria mutation follows a strict diet that avoids this amino acid, however, they remain normal and healthy.

Gene Regulation

The genome of a given organism contains thousands of genes, but not all these genes need to be active at any given moment. A gene is expressed when it is being transcribed into mRNA (and translated into protein), and there exist many cellular methods of controlling the expression of genes such that proteins are produced only when needed by the cell. Transcription factors are regulatory proteins that bind to the start of genes, either promoting or inhibiting the transcription of the gene.

Within the genome of Escherichia coli bacteria, for example, there exists a series of genes necessary for the synthesis of the amino acid tryptophan. However, when tryptophan is already available to the cell, these genes for tryptophan synthesis are no longer needed. The presence of tryptophan directly affects the activity of the genes—tryptophan molecules bind to the tryptophan repressor (a transcription factor), changing the repressor's structure such that the repressor binds to the genes. The tryptophan repressor blocks the transcription and expression of the genes, thereby creating negative feedback regulation of the tryptophan synthesis process. Differences in gene expression are especially clear within multicellular organisms, where cells all contain the same genome but have very different structures and behaviors due to the expression of different sets of genes.

All the cells in a multicellular organism derive from a single cell, differentiating into variant cell types in response to external and intercellular signals and gradually establishing different patterns of gene expression to create different behaviors. As no single gene is responsible for the development of structures within multicellular organisms, these patterns arise from the complex interactions between many cells.

Within eukaryotes there exist structural features of chromatin that influence the transcription of genes, often in the form of modifications to DNA and chromatin that are stably inherited by daughter cells These features are called "epigenetic" because they exist "on top" of the DNA sequence and retain inheritance from one cell generation to the next. Because of epigenetic features, different cell types grown within the same medium can retain very different properties. Although epigenetic features are generally dynamic over the course of development, some, like the phenomenon of paramutation, have multigenerational inheritance and exist as rare exceptions to the general rule of DNA as the basis for inheritance.

GENETIC CHANGE

During the process of DNA replication, errors occasionally occur in the polymerization of the second strand. These errors, called mutations, can have an impact on the phenotype of an organism, especially if they occur within the protein coding sequence of a gene. Error rates are usually very low—1 error in every 10-100 million bases—due to the "proofreading" ability of DNA polymerases (Without proofreading error rates are a thousand-fold higher; because many viruses rely on DNA and RNA polymerases that lack proofreading ability, they experience higher mutation rates.) Processes that increase the rate of changes in DNA are called mutagenic: mutagenic chemicals promote errors in DNA replication, often by interfering with the structure of base-pairing, while UV radiation induces mutations by causing damage to the DNA structure. Chemical damage to DNA occurs naturally as well, and cells use DNA repair mechanisms to repair mismatches and breaks in DNA—nevertheless, the repair sometimes fails to return the DNA to its original sequence.

In organisms that use chromosomal crossover to exchange DNA and recombine genes, errors in alignment during meiosis can also cause mutations Errors in crossover are especially likely when similar sequences cause partner chromosomes to adopt a mistaken alignment; this makes some regions in genomes more prone to mutating in this way. These errors create large structural changes in DNA sequence—duplications, inversions or deletions of entire regions, or the accidental exchanging of whole parts between different chromosomes (called translocation).

Natural Selection and Evolution

Mutations produce organisms with different genotypes, and those differences can result in different phenotypes. Many mutations have little effect on an organism's phenotype, health, and reproductive fitness. Mutations that do have an effect are often deleterious, but occasionally mutations are beneficial. Studies in the fly *Drosophila melanogaster* suggest that if a mutation changes a protein produced by a gene, this will probably be harmful, with about 70 percent of these mutations having damaging effects, and the remainder being either neutral or weakly beneficial.

An evolutionary tree of eukaryotic organisms, constructed by comparison of several orthologous gene sequencesPopulation genetics research studies the distributions of these genetic differences within populations and how the distributions change over time. Changes in the frequency of an allele in a population can be influenced by natural selection, where a given allele's higher rate of survival and reproduction causes it to become more frequent in the population over time. Genetic drift can also occur, where chance events lead to random changes in allele frequency.

Over many generations, the genomes of organisms can change, resulting in the phenomenon of evolution. Mutations and the selection for beneficial mutations can cause a species to evolve into forms that better survive their environment, a process called adaptation. New species are formed through the process of speciation, a process often caused by geographical separations that allow different populations to genetically diverge The application of genetic principles to the study of population biology and evolution is referred to as the modern synthesis.

As sequences diverge and change during the process of evolution, these differences between sequences can be used as a molecular clock to calculate the evolutionary distance between them. Genetic comparisons are generally considered the most accurate method of characterizing the relatedness between species, an improvement over the sometimes deceptive comparison of phenotypic characteristics. The evolutionary distances between species can be combined to form evolutionary trees - these trees represent the common descent and divergence of species over time, although they cannot represent the transfer of genetic material between unrelated species (known as horizontal gene transfer and most common in bacteria).

Model Organisms and Genetics

Although geneticists originally studied inheritance in a wide range of organisms, researchers began to specialize in studying the genetics of a particular subset of organisms. The fact that significant research already existed for a given organism would encourage new researchers to choose it for further study, and so eventually a few model organisms became the basis for most genetics research Common research topics in model organism genetics include the study of gene regulation and the involvement of genes in development and cancer.

Organisms were chosen, in part, for convenience—short generation times and easy genetic manipulation made some organisms popular genetics research tools. Widely used model organisms include the gut bacterium *Escherichia coli*, the plant *Arabidopsis thaliana*, baker's yeast (*Saccharomyces cerevisiae*), the nematode *Caenorhabditis elegans*, the common fruit fly (*Drosophila melanogaster*), and the common house mouse (*Mus musculus*).

Medical Genetics Research

When searching for an unknown gene that may be involved in a disease, researchers commonly use genetic linkage and genetic pedigree charts to find the location on the genome associated with the disease. At the population level, researchers take advantage of Mendelian randomization to look for locations in the genome that are associated with diseases, a technique especially useful for multigenic traits not clearly defined by a single gene. Once a candidate gene is found, further research is often done on the same gene (called an orthologous gene) in model organisms. In addition to studying genetic diseases, the increased availability of genotyping techniques has led to the field of pharmacogenetics—studying how genotype can affect drug responses. Although it is not an inherited disease, cancer is also considered a genetic disease.

The process of cancer development in the body is a combination of events. Mutations occasionally occur within cells in the body as they divide. While these mutations will not be inherited by any offspring, they can affect the behavior of cells, sometimes causing them to grow and divide more frequently. There are biological mechanisms that attempt to stop this process; signals are given to inappropriately dividing cells that should trigger cell death, but sometimes additional mutations occur that cause cells to ignore these messages. An internal process of natural selection occurs within the body and eventually mutations accumulate within cells to promote their own growth, creating a cancerous tumor that grows and invades various tissues of the body.

Research Techniques

DNA can be manipulated in the laboratory. Restriction enzymes are a commonly used enzyme that cuts DNA at specific sequences, producing predictable fragments of DNA The use of ligation enzymes allows these fragments to be reconnected, and by ligating fragments of DNA together from different sources, researchers can create recombinant DNA. Often associated with genetically modified organisms, recombinant DNA is commonly used in the context of plasmids - short circular DNA fragments with a few genes on them. By inserting plasmids into bacteria and growing those bacteria on plates of agar (to isolate clones of bacteria cells), researchers can clonally amplify the inserted fragment of DNA (a process known as molecular cloning). (Cloning can also refer to the creation of clonal organisms, through various techniques.)

The DNA can also be amplified using a procedure called the polymerase chain reaction (PCR) By using specific short sequences of DNA, PCR can isolate and exponentially amplify a targeted region of DNA. Because it can amplify from extremely small amounts of DNA, PCR is also often used to detect the presence of specific DNA sequences.

DNA Sequencing and Genomics

One of the most fundamental technologies developed to study genetics, DNA sequencing allows researchers to determine the sequence of nucleotides in DNA fragments. Developed in 1977 by Frederick Sanger and coworkers, chain-termination sequencing is now routinely used to sequence DNA fragments. With this technology, researchers have been able to study the molecular sequences associated with many human diseases. As sequencing has become less expensive and with the aid of computational tools, researchers have sequenced the genomes of many organisms by stitching together the sequences of many different fragments (a process called genome assembly). These technologies were used to sequence the human genome, leading to the completion of the Human Genome Project in 2003.

New high-throughput sequencing technologies are dramatically lowering the cost of DNA sequencing, with many researchers hoping to bring the cost of resequencing a human genome down to a thousand dollars. The large amount of sequences available has created the field of genomics, research that uses computational tools to search for and analyze patterns in the full genomes of organisms. Genomics can also be considered a subfield of bioinformatics, which uses computational approaches to analyze large sets of biological data.

Gene

A gene is the basic unit of heredity in a living organism. The field of genetics predates modern molecular biology, but it is now known that all living things depend on DNA to pass on their traits to offspring. Loosely speaking, a gene is a segment of genomic information that, taken as a whole, specifies a trait. The colloquial usage of the term gene often refers to the scientific concept of an allele.

The notion of a gene has evolved with the science of genetics, which began when Gregor Mendel noticed that biological variations were only inherited from parent organisms as specific, discrete traits. The biological entity responsible for defining traits was termed a "gene", but the biological basis for inheritance remained unknown until DNA was identified as the genetic material in the 1940s. All organisms have many genes corresponding to many different biological traits, some of which are immediately visible, such as eye color or number of limbs, and some of which are not, such as blood type or increased risk for certain diseases, or the thousands of basic biochemical processes which comprise life.

In cells, a gene is a portion of an organism's DNA which contains both "coding" sequences that determine what the gene does, and "non-coding" sequences that determine when the gene is active. When a gene is active, the coding and non-coding sequences are copied in a process called transcription, producing an RNA copy of the gene's information. This piece of RNA can then direct the synthesis of proteins via the genetic code. In other cases, the RNA is used directly, for example as part of the ribosome. The RNA may undergo special post-transcriptional processing steps required to convert it into a mature, functional form. These molecules resulting from gene expression, whether RNA or protein, are known as gene products, and are responsible for the development and functioning of all living things.

More technically, a gene is a locatable region of genomic sequence, corresponding to a unit of inheritance, and is associated with regulatory regions, transcribed regions and/or other functional sequence regions. The physical development and phenotype of organisms can be thought of as a product of genes interacting with each other and with the environment. A concise definition of a gene, taking into account complex patterns of regulation and transcription, genic conservation and non-coding RNA genes, has been proposed by Gerstein and co-workers "A gene is a union of genomic sequences encoding a coherent set of potentially overlapping functional products".

The existence of genes was first suggested by Gregor Mendel (1822-1884), who, in the 1860s, studied inheritance in peaplants and hypothesized a factor that conveys traits from parent to offspring. He spent over 10 years of his life on one experiment. Although he did not use the term gene, he explained his results in terms of inherited characteristics. Mendel was also the first to hypothesize independent assortment, the distinction between dominant and recessive traits, the distinction between a heterozygote and homozygote, and the difference between what would later be described as genotype and phenotype. Mendel's concept was given a name by Hugo de Vries in 1889, who, at that time probably unaware of Mendel's work, in his book Intracellular Pangenesis coined the term "pangen" for "the smallest particle [representing] one hereditary characteristic". Wilhelm Johannsen abbreviated this term to "gene" ("gen" in Danish and German) two decades later.

In the early 1900s, Mendel's work received renewed attention from scientists. In 1910, Thomas Hunt Morgan showed that genes reside on specific chromosomes. He later showed that genes occupy specific locations on the chromosome. With this knowledge, Morgan and his students began the first chromosomal map of the fruit fly *Drosophila*. In 1928, Frederick Griffith showed that genes could be transferred. In what is now known as Griffith's experiment, injections into a mouse of a deadly strain of bacteria that had been heat-killed transferred genetic information to a safe strain of the same bacteria, killing the mouse.

In 1941,George Wells Beadle and Edward Lawrie Tatum showed that mutations in genes caused errors in certain steps in metabolic pathways. This showed that specific genes code for specific proteins, leading to the "one gene, one enzyme" hypothesis Oswald Avery, Collin Macleod, and Maclyn McCarty showed in 1944 that DNA holds the gene's information In 1953, James D. Watson and Francis Crick demonstrated the molecular structure of DNA. Together, these discoveries established the central dogma of molecular biology, which states that proteins are translated from RNA which is transcribed from DNA. This dogma has since been shown to have exceptions, such as reverse transcription in retroviruses.

In 1972, Walter Fiers and his team at the Laboratory of Molecular Biology of the University of Ghent (Ghent, Belgium) were the first to determine the sequence of a gene: the gene for Bacteriophage MS2 coat protein Richard J. Roberts and Phillip Sharp discovered in 1977 that genes can be split into segments. This leads to the idea that one gene can make several proteins. In 2003-2006), biological results let the notion of gene appear more slippery. In particular, genes do not seem to sit side by side on DNA like discrete beads. Instead, regions of the DNA producing distinct proteins may overlap, so that the idea emerges that "genes are one long continuum".

Mendelian Inheritance and Classical Genetics

Darwin used the term Gemmule to describe a microscopic unit of inheritance, and what would later become known as 'Chromosomes' had been observed separating out during cell division by Wilhelm Hofmeister as early as 1848. The idea that chromosomes were the carriers of inheritance was expressed in 1883 by Wilhelm Roux. The modern conception of the gene originated with work by Gregor Mendel, a 19th century Augustinian monk who systematically studied heredity in pea plants. Mendel's work was the first to illustrate particulate inheritance, or the theory that inherited traits are passed from one generation to the next in discrete units that interact in well-defined ways. Danish botanist Wilhelm Johannsen coined the word "gene" in 1909 to describe these fundamental physical and functional units of heredity,[9] while the related word genetics was first used by William Bateson in 1905 The word was derived from Hugo De Vries' 1889 term pangen for the same concept, itself a derivative of the word pangenesis coined by Darwin (1868). The word pangenesis is made from the Greek words pan (a prefix meaning "whole", "encompassing") and genesis ("birth") or genos ("origin").

Crossing between two pea plants heterozygous for purple (B, dominant) and white (b, recessive) blossoms. According to the theory of Mendelian inheritance, variations in phenotype - the observable physical and behavioural characteristics of an organism - are due to variations in genotype, or the organism's particular set of genes, each of which specifies a particular trait. Different forms of a gene, which may give rise to different phenotypes, are known as alleles. Organisms such as the pea plants Mendel worked on, along with many plants and animals, have two alleles for each trait, one inherited from each parent. Alleles may be dominant or recessive; dominant alleles give rise to their corresponding phenotypes when paired with any other allele for the same trait, while recessive alleles give rise to their corresponding phenotype only when paired with another copy of the same allele. For example, if the allele specifying tall stems in pea plants is dominant over the allele specifying short stems, then pea plants that inherit one tall allele from one parent and one short allele from the other parent will also have tall stems. Mendel's work found that alleles assort independently in the production of gametes, or germ cells, ensuring variation in the next generation.

Prior to Mendel's work, the dominant theory of heredity was one of blending inheritance, which proposes that the traits of the parents blend or mix in a smooth, continuous gradient in the offspring. Although Mendel's work was largely unrecognized after its first publication in 1866, it was rediscovered in 1900 by three European scientists, Hugo de Vries, Carl Correns, and Erich von Tschermak, who had reached similar conclusions from their own research. However, these scientists were not yet aware of the identity of the 'discrete units' on which genetic material resides.

A series of subsequent discoveries led to the realization decades later that chromosomes within cells are the carriers of genetic material, and that they are made of DNA (deoxyribonucleic acid), a polymeric molecule found in all cells on which the 'discrete units' of Mendelian inheritance are encoded. The modern study of genetics at the level of DNA is known as molecular genetics and the synthesis of molecular genetics with traditional Darwinian evolution is known as the modern evolutionary synthesis.

Physical Definitions

The vast majority of living organisms encode their genes in long strands of DNA. DNA consists of a chain made from four types of nucleotide subunits: adenine, cytosine, guanine, and thymine. Each nucleotide subunit consists of three components: a phosphate group, a deoxyribose sugar ring, and a nucleobase. Thus, nucleotides in DNA or RNA are typically called 'bases'; consequently they are commonly referred to simply by their purine or pyrimidine original base components adenine, cytosine, guanine, thymine. Adenine and guanine are purines and cytosine and thymine are pyrimidines. The most common form of DNA in a cell is in a double helix structure, in which two individual DNA strands twist around each other in a right-handed spiral. In this structure, the base pairing rules specify that guanine pairs with cytosine and adenine pairs with thymine (each pair contains one purine and one pyrimidine). The base pairing between guanine and cytosine forms three hydrogen bonds, while the base pairing between adenine and thymine forms two hydrogen bonds. The two strands in a double helix must therefore be complementary, that is, their bases must align such that the adenines of one strand are paired with the thymines of the other strand, and so on.

Due to the chemical composition of the pentose residues of the bases, DNA strands have directionality. One end of a DNA polymer contains an exposed hydroxyl group on the deoxyribose, this is known as the 3' end of the molecule. The other end contains an exposed phosphate group, this is the 5' end. The directionality of DNA is vitally important to many cellular processes, since double helices are necessarily directional (a strand running 5'-3' pairs with a complementary strand running 3'-5') and processes such as DNA replication occur in only one direction. All nucleic acid synthesis in a cell occurs in the 5'-3' direction, because new monomers are added via a dehydration reaction that uses the exposed 3' hydroxyl as a nucleophile.

The expression of genes encoded in DNA begins by transcribing the gene into RNA, a second type of nucleic acid that is very similar to DNA, but whose monomers contain the sugar ribose rather than deoxyribose. RNA also contains the base uracil in place of thymine. RNA molecules are less stable than DNA and are typically single-stranded. Genes that encode proteins are composed of a series of three-nucleotide sequences called codons, which serve as the “words” in the genetic “language”. The genetic code specifies the correspondence during protein translation between codons and amino acids. The genetic code is nearly the same for all known organisms.

RNA Genes and Genomes

In some cases, RNA is an intermediate product in the process of manufacturing proteins from genes. However, for other gene sequences, the RNA molecules are the actual functional products. For example, RNAs known as ribozymes are capable of enzymatic function, and miRNAs have a regulatory role. The DNA sequences from which such RNAs are transcribed are known as RNA genes.

Some viruses store their entire genomes in the form of RNA, and contain no DNA at all. Because they use RNA to store genes, their cellular hosts may synthesize their proteins as soon as they are infected and without the delay in waiting for transcription. On the other hand, RNA retroviruses, such as HIV, require the reverse transcription of their genome from RNA into DNA before their proteins can be synthesized. In 2006, French researchers came across a puzzling example of RNA-mediated inheritance in mouse. Mice with a loss-of-function mutation in the gene Kit have white tails. Offspring of these mutants can have white tails despite having only normal Kit genes. The research team traced this effect back to mutated Kit RNA While RNA is common as genetic storage material in viruses, in mammals in particular RNA inheritance has been observed very rarely.

Functional Structure of a Gene

The pre-mRNA is then spliced into messenger RNA (mRNA) which is later translated into protein.All genes have regulatory regions in addition to regions that explicitly code for a protein or RNA product. A regulatory region shared by almost all genes is known as the promoter, which provides a position that is recognized by the transcription machinery when a gene is about to be transcribed and expressed. A gene can have more than one promoter, resulting in RNAs that differ in how far they extend in the 5' end Although promoter regions have a consensus sequence that is the most common sequence at this position, some genes have "strong" promoters that bind the transcription machinery well, and others have "weak" promoters that bind poorly. These weak promoters usually permit a lower rate of transcription than the strong promoters, because the transcription machinery binds to them and initiates transcription less frequently. Other possible regulatory regions include enhancers, which can compensate for a weak promoter. Most regulatory regions are "upstream" — that is, before or toward the 5' end of the transcription initiation site. Eukaryotic promoter regions are much more complex and difficult to identify than prokaryotic promoters.

Many prokaryotic genes are organized into operons, or groups of genes whose products have related functions and which are transcribed as a unit. By contrast, eukaryotic genes are transcribed only one at a time, but may include long stretches of DNA called introns which are transcribed but never translated into protein (they are spliced out before translation). Splicing can also occur in prokaryotic genes, but is less common than in eukaryotes

Chromosomes

The total complement of genes in an organism or cell is known as its genome, which may be stored on one or more chromosomes; the region of the chromosome at which a particular gene is located is called its locus. A chromosome consists of a single, very long DNA helix on which thousands of genes are encoded. Prokaryotes - bacteria and archaea - typically store their genomes on a single large, circular chromosome, sometimes supplemented by additional small circles of DNA called plasmids, which usually encode only a few genes and are easily transferable between individuals. For example, the genes for antibiotic resistance are usually encoded on bacterial plasmids and can be passed between individual cells, even those of different species, via horizontal gene transfer. Although some simple eukaryotes also possess plasmids with small numbers of genes, the majority of eukaryotic genes are stored on multiple linear chromosomes, which are packed within the nucleus in complex with storage proteins called histones. The manner in which DNA is stored on the histone, as well as chemical modifications of the histone itself, are regulatory mechanisms governing whether a particular region of DNA is accessible for gene expression. The ends of eukaryotic chromosomes are capped by long stretches of repetitive sequences called telomeres, which do not code for any gene product but are present to prevent degradation of coding and regulatory regions during DNA replication. The length of the telomeres tends to decrease each time the genome is replicated in preparation for cell division; the loss of telomeres has been proposed as an explanation for cellular senescence, or the loss of the ability to divide, and by extension for the aging process in organisms.

While the chromosomes of prokaryotes are relatively gene-dense, those of eukaryotes often contain so-called "junk DNA", or regions of DNA that serve no obvious function. Simple single-celled eukaryotes have relatively small amounts of such DNA, while the genomes of complex multicellular organisms, including humans, contain an absolute majority of DNA without an identified function. However it now appears that, although protein-coding DNA makes up barely 2% of the human genome, about 80% of the bases in the genome may be being expressed, so the term "junk DNA" may be a misnomer.

Gene Expression

In all organisms, there are two major steps separating a protein-coding gene from its protein: first, the DNA on which the gene resides must be transcribed from DNA to messenger RNA (mRNA), and second, it must be translated from mRNA to protein. RNA-coding genes must still go through the first step, but are not translated into protein. The process of producing a biologically functional molecule of either RNA or protein is called gene expression, and the resulting molecule itself is called a gene product.

The genetic code is the set of rules by which a gene is translated into a functional protein. Each gene consists of a specific sequence of nucleotides encoded in a DNA (or sometimes RNA) strand; a correspondence between nucleotides, the basic building blocks of genetic material, and amino acids, the basic building blocks of proteins, must be established for genes to be successfully translated into functional proteins.

Sets of three nucleotides, known as codons, each correspond to a specific amino acid or to a signal; three codons are known as "stop codons" and, instead of specifying a new amino acid, alert the translation machinery that the end of the gene has been reached. There are 64 possible codons (four possible nucleotides at each of three positions, hence 43 possible codons) and only 20 standard amino acids; hence the code is redundant and multiple codons can specify the same amino acid. The correspondence between codons and amino acids is nearly universal among all known living organisms.

Transcription

The process of genetic transcription produces a single-stranded RNA molecule known as messenger RNA, whose nucleotide sequence is complementary to the DNA from which it was transcribed. The DNA strand whose sequence matches that of the RNA is known as the coding strand and the strand from which the RNA was synthesized is the template strand. Transcription is performed by an enzyme called an RNA polymerase, which reads the template strand in the 3' to 5' direction and synthesizes the RNA from 5' to 3'. To initiate transcription, the polymerase first recognizes and binds a promoter region of the gene. Thus a major mechanism of gene regulation is the blocking or sequestering of the promoter region, either by tight binding by repressor molecules that physically block the polymerase, or by organizing the DNA so that the promoter region is not accessible.

In prokaryotes, transcription occurs in the cytoplasm; for very long transcripts, translation may begin at the 5' end of the RNA while the 3' end is still being transcribed. In eukaryotes, transcription necessarily occurs in the nucleus, where the cell's DNA is sequestered; the RNA molecule produced by the polymerase is known as the primary transcript and must undergo post-transcriptional modifications before being exported to the cytoplasm for translation. The splicing of introns present within the transcribed region is a modification unique to eukaryotes; alternative splicing mechanisms can result in mature transcripts from the same gene having different sequences and thus coding for different proteins. This is a major form of regulation in eukaryotic cells.

TRANSLATION

Translation is the process by which a mature mRNA molecule is used as a template for synthesizing a new protein. Translation is carried out by ribosomes, large complexes of RNA and protein responsible for carrying out the chemical reactions to add new amino acids to a growing polypeptide chain by the formation of peptide bonds. The genetic code is read three nucleotides at a time, in units called codons, via interactions with specialized RNA molecules called transfer RNA (tRNA). Each tRNA has three unpaired bases known as the anticodon that are complementary to the codon it reads; the tRNA is also covalently attached to the amino acid specified by the complementary codon. When the tRNA binds to its complementary codon in an mRNA strand, the ribosome ligates its amino acid cargo to the new polypeptide chain, which is synthesized from amino terminus to carboxyl terminus. During and after its synthesis, the new protein must fold to its active three-dimensional structure before it can carry out its cellular function.

DNA Replication and Inheritance

The growth, development, and reproduction of organisms relies on cell division, or the process by which a single cell divides into two usually identical daughter cells. This requires first making a duplicate copy of every gene in the genome in a process called DNA replication. The copies are made by specialized enzymes known as DNA polymerases, which "read" one strand of the double-helical DNA, known as the template strand, and synthesize a new complementary strand. Because the DNA double helix is held together by base pairing, the sequence of one strand completely specifies the sequence of its complement; hence only one strand needs to be read by the enzyme to produce a faithful copy. The process of DNA replication is semiconservative; that is, the copy of the genome inherited by each daughter cell contains one original and one newly synthesized strand of DNA.

After DNA replication is complete, the cell must physically separate the two copies of the genome and divide into two distinct membrane-bound cells. In prokaryotes - bacteria and archaea - this usually occurs via a relatively simple process called binary fission, in which each circular genome attaches to the cell membrane and is separated into the daughter cells as the membrane invaginates to split the cytoplasm into two membrane-bound portions. Binary fission is extremely fast compared to the rates of cell division in eukaryotes. Eukaryotic cell division is a more complex process known as the cell cycle; DNA replication occurs during a phase of this cycle known as S phase, while the process of segregating chromosomes and splitting the cytoplasm occurs during M phase. In many single-celled eukaryotes such as yeast, reproduction by budding is common, which results in asymmetrical portions of cytoplasm in the two daughter cells.

Molecular Inheritance

The duplication and transmission of genetic material from one generation of cells to the next is the basis for molecular inheritance, and the link between the classical and molecular pictures of genes. Organisms inherit the characteristics of their parents because the cells of the offspring contain copies of the genes in their parents' cells. In asexually reproducing organisms, the offspring will be a genetic copy or clone of the parent organism. In sexually reproducing organisms, a specialized form of cell division called meiosis produces cells called gametes or germ cells that are haploid, or contain only one copy of each gene. The gametes produced by females are called eggs or ova, and those produced by males are called sperm. Two gametes fuse to form a fertilized egg, a single cell that once again has a diploid number of genes - each with one copy from the mother and one copy from the father.

During the process of meiotic cell division, an event called genetic recombination or crossing-over can sometimes occur, in which a length of DNA on one chromatid is swapped with a length of DNA on the corresponding sister chromatid. This has no effect if the alleles on the chromatids are the same, but results in reassortment of otherwise linked alleles if they are different. The Mendelian principle of independent assortment asserts that each of a parent's two genes for each trait will sort independently into gametes; which allele an organism inherits for one trait is unrelated to which allele it inherits for another trait. This is in fact only true for genes that do not reside on the same chromosome, or are located very far from one another on the same chromosome. The closer two genes lie on the same chromosome, the more closely they will be associated in gametes and the more often they will appear together; genes that are very close are essentially never separated because it is extremely unlikely that a crossover point will occur between them. This is known as genetic linkage.

Mutation

The DNA replication is for the most part extremely accurate, with an error rate per site of around 10-6 to 10-10 in eukaryotes. Rare, spontaneous alterations in the base sequence of a particular gene arise from a number of sources, such as errors in DNA replication and the aftermath of DNA damage. These errors are called mutations. The cell contains many DNA repair mechanisms for preventing mutations and maintaining the integrity of the genome; however, in some cases—such as breaks in both DNA strands of a chromosome—repairing the physical damage to the molecule is a higher priority than producing an exact copy. Due to the degeneracy of the genetic code, some mutations in protein-coding genes are silent, or produce no change in the amino acid sequence of the protein for which they code; for example, the codons UCU and UUC both code for serine, so the U?C mutation has no effect on the protein. Mutations that do have phenotypic effects are most often neutral or deleterious to the organism, but sometimes they confer benefits to the organism's fitness.

Mutations propagated to the next generation lead to variations within a species' population. Variants of a single gene are known as alleles, and differences in alleles may give rise to differences in traits. Although it is rare for the variants in a single gene to have clearly distinguishable phenotypic effects, certain well-defined traits are in fact controlled by single genetic loci. A gene's most common allele is called the wild type allele, and rare alleles are called mutants. However, this does not imply that the wild-type allele is the ancestor from which the mutants are descended.

Genome

Chromosomal Organisation

The total complement of genes in an organism or cell is known as its genome. In prokaryotes, the vast majority of genes are located on a single chromosome of circular DNA, while eukaryotes usually possess multiple individual linear DNA helices packed into dense DNA-protein complexes called chromosomes. Genes that appear together on one chromosomes of one species may appear on separate chromosomes in another species. Many species carry more than one copy of their genome within each of their somatic cells. Cells or organisms with only one copy of each chromosome are called haploid; those with two copies are called diploid; and those with more than two copies are called polyploid. The copies of genes on the chromosomes are not necessarily identical. In sexually reproducing organisms, one copy is normally inherited from each parent.

Composition of the Genome

Typical numbers of genes and size of genomes vary widely among organisms, even those that are fairly closely evolutionarily related. Although it was believed before the completion of the Human Genome Project that the human genome would contain many more genes than simpler animals such as mice or fruit flies, the completion of the project has revealed that the human genome has an unexpectedly low gene density Estimates of the number of genes in a genome are difficult to compile because they depend on gene finding algorithms that search for patterns resembling those present in known genes, such as open reading frames, promoter regions with sequences resembling the consensus promoter sequence, and related regulatory regions such as TATA boxes in eukaryotes. Gene finding is less reliable in eukaryotic than in prokaryotic genomes due to the presence of non-coding DNA such as introns and pseudogenes. Computational gene finding methods are still significantly more reliable than earlier techniques that required mapping the locations of specific mutations that gave rise to distinguishable alleles.

In most eukaryotic species, very little of the DNA in the genome encodes proteins, and the genes may be separated by vast regions of non-coding DNA, much of which has been labeled "junk DNA" due to its apparent lack of function in the modern organism. A commonly studied type of "junk DNA" is the pseudogenes, or region of non-coding DNA that resembles expressed genes but usually lacks appropriate promoters and other control sequences; such regions are hypothesized to be the results of gene duplication events in a lineage's evolutionary past. Moreover, the genes are often fragmented internally by non-coding sequences called introns, which can be many times longer than the coding sequence but are spliced during post-transcriptional modification of pre-mRNA.

Genetic and Genomic Nomenclature

Gene nomenclature has been established by the HUGO Gene Nomenclature Committee (HGNC) for each known human gene in the form of an approved gene name and symbol (short-form abbreviation). All approved symbols are stored in the HGNC Database. Each symbol is unique and each gene is only given one approved gene symbol. It is necessary to provide a unique symbol for each gene so that people can talk about them. This also facilitates electronic data retrieval from publications. In preference each symbol maintains parallel construction in different members of a gene family and can be used in other species, especially the mouse.

Evolutionary Concept of a Gene

George C. Williams first explicitly advocated the gene-centric view of evolution in his 1966 book *Adaptation and Natural Selection*. He proposed an evolutionary concept of gene to be used when we are talking about natural selection favoring some genes. The definition is: "that which segregates and recombines with appreciable frequency." According to this definition, even an asexual genome could be considered a gene, insofar that it have an appreciable permanency through many generations.

The difference is: the molecular gene transcribes as a unit, and the evolutionary gene inherits as a unit.

Richard Dawkins' books *The Selfish Gene* (1976) and The *Extended Phenotype* (1982) defended the idea that the gene is the only replicator in living systems. This means that only genes transmit their structure largely intact and are potentially immortal in the form of copies. So, genes should be the unit of selection. In The Selfish Gene Dawkins attempts to redefine the word 'gene' to mean "an inheritable unit" instead of the generally accepted definition of "a section of DNA coding for a particular protein". In River Out of Eden, Dawkins further refined the idea of gene-centric selection by describing life as a river of compatible genes flowing through geological time. Scoop up a bucket of genes from the river of genes, and we have an organism serving as temporary bodies or survival machines. A river of genes may fork into two branches representing two non-interbreeding species as a result of geographical separation.

Gene Targeting and Implications

Gene targeting is commonly referred to techniques for altering or disrupting mouse genes and provides the mouse models for studying the roles of individual genes in embryonic development, human disorders, aging and diseases. The mouse models, where one or more of its genes are deactivated or made inoperable, are called knockout mice. Since the first reports in which homologous recombination in embryonic stem cells was used to generate gene-targeted mice, gene targeting has proven to be a powerful means of precisely manipulating the mammalian genome, producing at least ten thousand mutant mouse strains and it is now possible to introduce mutations that can be activated at specific time points, or in specific cells or organs, both during development and in the adult animal.

Gene targeting strategies have been expanded to all kinds of modifications, including point mutations, isoform deletions, mutant allele correction, large pieces of chromosomal DNA insertion and deletion, tissue specific disruption combined with spatial and temporal regulation and so on. It is predicted that the ability to generate mouse models with predictable phenotypes will have a major impact on studies of all phases of development, immunology, neurobiology, oncology, physiology, metabolism, and human diseases. Gene targeting is also in theory applicable to species from which totipotent embryonic stem cells can be established, and therefore may offer a potential to the improvement of domestic animals and plants.

Changing Concept

The concept of the gene has changed considerably. From the original definition of a "unit of inheritance", the term evolved to mean a DNA-based unit that can exert its effects on the organism through RNA or protein products. It was also previously believed that one gene makes one protein; this concept was overthrow.

The definition of a gene is still changing. The first cases of RNA-based inheritance have been discovered in mammals. Evidence is also accumulating that the control regions of a gene do not necessarily have to be close to the coding sequence on the linear molecule or even on the same chromosome. Spilianakis and colleagues discovered that the promoter region of the interferon-gamma gene on chromosome 10 and the regulatory regions of the T(H)2 cytokine locus on chromosome 11 come into close proximity in the nucleus possibly to be jointly regulated.

The concept that genes are clearly delimited is also being eroded. There is evidence for fused proteins stemming from two adjacent genes that can produce two separate protein products. While it is not clear whether these fusion proteins are functional, the phenomena is more frequent than previously though Even more ground-breaking than the discovery of fused genes is the observation that some proteins can be composed of exons from far away regions and even different chromosomes

This new data has led to an updated, and probably tentative, definition of a gene as “a union of genomic sequences encoding a coherent set of potentially overlapping functional products. This new definition categorizes genes by functional products”, whether they be proteins or RNA, rather than specific DNA loci; all regulatory elements of DNA are therefore classified as gene-associated regions.

Chapter–3

EVOLUTION

INTRODUCTION

In biology, evolution is change in the inherited traits of a population of organisms from one generation to the next. These changes are caused by a combination of three main processes: variation, reproduction, and selection. Genes that are passed on to an organism's offspring produce the inherited traits that are the basis of evolution. These traits vary within populations, with organisms showing heritable differences in their traits. When organisms reproduce, their offspring may have new or altered traits. These new traits arise in two main ways: either from mutations in genes, or from the transfer of genes between populations and between species. In species that reproduce sexually, new combinations of genes are also produced by genetic recombination, which can increase variation between organisms. Evolution occurs when these heritable differences become more common or rare in a population.

Two major mechanisms drive evolution. The first is natural selection, a process causing heritable traits that are helpful for survival and reproduction to become more common in a population, and harmful traits to become more rare. This occurs because individuals with advantageous traits are more likely to reproduce, so that more individuals in the next generation inherit these traits. Over many generations, adaptations occur through a combination of successive, small, random changes in traits, and natural selection of those variants best-suited for their environment. The second major mechanism is genetic drift, an independent process that produces random changes in the frequency of traits in a population. Genetic drift results from the role probability plays in whether a given trait will be passed on as individuals survive and reproduce. Though the changes produced in any one generation by drift and selection are small, differences accumulate with each subsequent generation and can, over time, cause substantial changes in the organisms. This process can culminate in the emergence of new species. Indeed, the similarities between organisms suggest that all known species are descended from a common ancestor (or ancestral gene pool) through this process of gradual divergence.

Evolutionary biology documents the fact that evolution occurs, and also develops and tests theories that explain its causes. Studies of the fossil record and the diversity of living organisms had convinced most scientists by the mid-nineteenth century that species changed over time. However, the mechanism driving these changes remained unclear until the 1859 publication of Charles Darwin's On the Origin of Species, detailing the theory of evolution by natural selection Darwin's work soon led to overwhelming acceptance of evolution within the scientific community In the 1930s, Darwinian natural selection was combined with Mendelian inheritance to form the modern evolutionary synthesis in which the connection between the units of evolution (genes) and the mechanism of evolution (natural selection) was made. This powerful explanatory and predictive theory directs research by constantly raising new questions, and it has become the central organizing principle of modern biology, providing a unifying explanation for the diversity of life on Earth.

Evolution in organisms occurs through changes in heritable traits – particular characteristics of an organism. In humans, for example, eye color is an inherited characteristic, which individuals can inherit from one of their parents. Inherited traits are controlled by genes and the complete set of genes within an organism's genome is called its genotype.

The complete set of observable traits that make up the structure and behavior of an organism is called its phenotype. These traits come from the interaction of its genotype with the environment As a result, not every aspect of an organism's phenotype is inherited. Suntanned skin results from the interaction between a person's genotype and sunlight; thus, suntans are not passed on to people's children. However, people have different responses to sunlight, arising from differences in their genotype; a striking example is individuals with the inherited trait of albinism, who do not tan and are highly sensitive to sunburn.

Heritable traits are propagated between generations via DNA, a molecule which is capable of encoding genetic information. The DNA is a polymer composed of four types of bases. The sequence of bases along a particular DNA molecule specify the genetic information, in a manner akin to a sequence of letters specifying a text or a sequence of bits specifying a computer programme. Portions of a DNA molecule that specify a single functional unit are called genes; different genes have different sequences of bases. Within cells, the long strands of DNA associate with proteins to form condensed structures called chromosomes. A specific location within a chromosome is known as a locus. If the DNA sequence at a locus varies between individuals, the different forms of this sequence are called alleles. The DNA sequences can change through mutations, producing new alleles. If a mutation occurs within a gene, the new allele may affect the trait that the gene controls, altering the phenotype of the organism. However, while this simple correspondence between an allele and a trait works in some cases, most traits are more complex and are controlled by multiple interacting genes

Variation

An individual organism's phenotype results from both its genotype and the influence from the environment it has lived in. A substantial part of the variation in phenotypes in a population is caused by the differences between their genotypes. The modern evolutionary synthesis defines evolution as the change over time in this genetic variation. The frequency of one particular allele will fluctuate, becoming more or less prevalent relative to other forms of that gene. Evolutionary forces act by driving these changes in allele frequency in one direction or another. Variation disappears when an allele reaches the point of fixation-when it either disappears from the population or replaces the ancestral allele entirely.

Variation comes from mutations in genetic material, migration between populations (gene flow), and the reshuffling of genes through sexual reproduction. Variation also comes from exchanges of genes between different species; for example, through horizontal gene transfer in bacteria, and hybridization in plants Despite the constant introduction of variation through these processes, most of the genome of a species is identical in all individuals of that species However, even relatively small changes in genotype can lead to dramatic changes in phenotype: chimpanzees and humans differ in only about 5% of their genome.

Mutation

Application of part of a chromosomeGenetic variation comes from random mutations that occur in the genomes of organisms. Mutations are changes in the DNA sequence of a cell's genome and are caused by radiation, viruses, transposons and mutagenic chemicals, as well as errors that occur during meiosis or DNA replication. These mutagens produce several different types of change in DNA sequences; these can either have no effect, alter the product of a gene, or prevent the gene from functioning. Studies in the fly *Drosophila melanogaster* suggest that if a mutation changes a protein produced by a gene, this will probably be harmful, with about 70 per cent of these mutations having damaging effects, and the remainder being either neutral or weakly beneficial. Due to the damaging effects that mutations can have on cells, organisms have evolved mechanisms such as DNA repair to remove mutations. Therefore, the optimal mutation rate for a species is a trade-off between costs of a high mutation rate, such as deleterious mutations, and the metabolic costs of maintaining systems to reduce the mutation rate, such as DNA repair enzymes Some species such as retroviruses have such high mutation rates that most of their offspring will possess a mutated gene. Such rapid mutation may have been selected so that these viruses can constantly and rapidly evolve, and thus evade the responses of the human immune system.

Mutations can involve large sections of DNA becoming duplicated, which is a major source of raw material for evolving new genes, with tens to hundreds of genes duplicated in animal genomes every million years. Most genes belong to larger families of genes of shared ancestry. Novel genes are produced by several methods, commonly through the duplication and mutation of an ancestral gene, or by recombining parts of different genes to form new combinations with new functions. For example, the human eye uses four genes to make structures that sense light: three for color vision and one for night vision; all four arose from a single ancestral gene. An advantage of duplicating a gene (or even an entire genome) is that overlapping or redundant functions in multiple genes allows alleles to be retained that would otherwise be harmful, thus increasing genetic diversity.

Changes in chromosome number may involve even larger mutations, where segments of the DNA within chromosomes break and then rearrange. For example, two chromosomes in the Homo genus fused to produce human chromosome 2; this fusion did not occur in the lineage of the other apes, and they retain these separate chromosomes. In evolution, the most important role of such chromosomal rearrangements may be to accelerate the divergence of a population into new species by making populations less likely to interbreed, and thereby preserving genetic differences between these populations.

Sequences of DNA that can move about the genome, such as transposons, make up a major fraction of the genetic material of plants and animals, and may have been important in the evolution of genomes For example, more than a million copies of the Alu sequence are present in the human genome, and these sequences have now been recruited to perform functions such as regulating gene expression. Another effect of these mobile DNA sequences is that when they move within a genome, they can mutate or delete existing genes and thereby produce genetic diversity.

RECOMBINATION

In asexual organisms, genes are inherited together, or linked, as they cannot mix with genes in other organisms during reproduction. However, the offspring of sexual organisms contain random mixtures of their parents' chromosomes that are produced through independent assortment. In the related process of genetic recombination, sexual organisms can also exchange DNA between two matching chromosomes. Recombination and reassortment do not alter allele frequencies, but instead change which alleles are associated with each other, producing offspring with new combinations of alleles. While this process increases the variation in any individual's offspring, genetic mixing can be predicted to either have no effect, increase, or decrease the genetic variation in the population, depending on how the various alleles in the population are distributed. For example, if two alleles are randomly distributed in a population, then sex will have no effect on variation; however, if two alleles tend to be found as a pair, then genetic mixing will even out this non-random distribution and over time make the organisms in the population more similar to each other The overall effect of sex on natural variation remains unclear, but recent research suggests that sex usually increases genetic variation and may increase the rate of evolution.

Recombination allows even alleles that are close together in a strand of DNA to be inherited independently. However, the rate of recombination is low, since in humans in a stretch of DNA one million base pairs long there is about a one in a hundred chance of a recombination event occurring per generation. As a result, genes close together on a chromosome may not always be shuffled away from each other, and genes that are close together tend to be inherited together. This tendency is measured by finding how often two alleles of different genes occur together, which is called their linkage disequilibrium. A set of alleles that is usually inherited in a group is called a haplotype.

Sexual reproduction helps to remove harmful mutations and retain beneficial mutations. Consequently, when alleles cannot be separated by recombination – such as in mammalian Y chromosomes, which pass intact from fathers to sons – harmful mutations accumulate. In addition, recombination and reassortment can produce individuals with new and advantageous gene combinations. These positive effects are balanced by the fact that this process can cause mutations and separate beneficial combinations of genes.

POPULATION GENETICS

From a genetic viewpoint, evolution is a generation-to-generation change in the frequencies of alleles within a population that shares a common gene pool. A population is a localized group of individuals belonging to the same species. For example, all of the moths of the same species living in an isolated forest represent a population. A single gene in this population may have several alternate forms, which account for variations between the phenotypes of the organisms. An example might be a gene for coloration in moths that has two alleles: black and white. A gene pool is the complete set of alleles in a single population, so each allele occurs a certain number of times in a gene pool. The fraction of genes within the gene pool that are a particular allele is called the allele frequency. Evolution occurs when there are changes in the frequencies of alleles within a population of interbreeding organisms; for example the allele for black color in a population of moths becoming more common.

To understand the mechanisms that cause a population to evolve, it is useful to consider what conditions are required for a population not to evolve. The Hardy-Weinberg principle states that the frequencies of alleles (variations in a gene) in a sufficiently large population will remain constant if the only forces acting on that population are the random reshuffling of alleles during the formation of the sperm or egg, and the random combination of the alleles in these sex cells during fertilization. Such a population is said to be in Hardy-Weinberg equilibrium - it is not evolving.

MECHANISMS

The main mechanisms for producing evolutionary change are natural selection, genetic drift, and gene flow. Natural selection favors genes that improve capacity for survival and reproduction. Genetic drift is random change in the frequency of alleles, caused by the random sampling of a generation's genes during reproduction. Gene flow is the transfer of genes within and between populations. The relative importance of natural selection and genetic drift in a population varies depending on the strength of the selection and the effective population size, which is the number of individuals capable of breeding. Natural selection usually predominates in large populations, while genetic drift dominates in small populations. The dominance of genetic drift in small populations can even lead to the fixation of slightly deleterious mutations As a result, changing population size can dramatically influence the course of evolution. Population bottlenecks, where the population shrinks temporarily and therefore loses genetic variation, result in a more uniform population. Bottlenecks also result from alterations in gene flow such as decreased migration, expansions into new habitats, or population subdivision.

Natural Selection

Natural selection is the process by which genetic mutations that enhance reproduction become, and remain, more common in successive generations of a population. It has often been called a "self-evident" mechanism because it necessarily follows from three simple facts:

Heritable variation exists within populations of organisms. Organisms produce more offspring than can survive. These offspring vary in their ability to survive and reproduce.

These conditions produce competition between organisms for survival and reproduction. Consequently, organisms with traits that give them an advantage over their competitors pass these advantageous traits on, while traits that do not confer an advantage are not passed on to the next generation.

The central concept of natural selection is the evolutionary fitness of an organism. This measures the organism's genetic contribution to the next generation. However, this is not the same as the total number of offspring: instead fitness measures the proportion of subsequent generations that carry an organism's genes. Consequently, if an allele increases fitness more than the other alleles of that gene, then with each generation this allele will become more common within the population. These traits are said to be "selected for". Examples of traits that can increase fitness are enhanced survival, and increased fecundity. Conversely, the lower fitness caused by having a less beneficial or deleterious allele results in this allele becoming rarer — they are "selected against". Importantly, the fitness of an allele is not a fixed characteristic, if the environment changes, previously neutral or harmful traits may become beneficial and previously beneficial traits become harmful.

Natural selection within a population for a trait that can vary across a range of values, such as height, can be categorized into three different types. The first is directional selection, which is a shift in the average value of a trait over time — for example organisms slowly getting taller. Secondly, disruptive selection is selection for extreme trait values and often results in two different values becoming most common, with selection against the average value. This would be when either short or tall organisms had an advantage, but not those of medium height. Finally, in stabilizing selection there is selection against extreme trait values on both ends, which causes a decrease in variance around the average value This would, for example, cause organisms to slowly become all the same height.

A special case of natural selection is sexual selection, which is selection for any trait that increases mating success by increasing the attractiveness of an organism to potential mates. Traits that evolved through sexual selection are particularly prominent in males of some animal species, despite traits such as cumbersome antlers, mating calls or bright colours that attract predators, decreasing the survival of individual males This survival disadvantage is balanced by higher reproductive success in males that show these hard to fake, sexually selected traits.

An active area of research is the unit of selection, with natural selection being proposed to work at the level of genes, cells, individual organisms, groups of organisms and even species. None of these models are mutually exclusive and selection may act on multiple levels simultaneously. Below the level of the individual, genes called transposons try to copy themselves throughout the genome. Selection at a level above the individual, such as group selection, may allow the evolution of co-operation, as discussed below.

GENETIC DRIFT

Genetic drift is the change in allele frequency from one generation to the next that occurs because alleles in offspring are a random sample of those in the parents, as well as from the role that chance plays in determining whether a given individual will survive and reproduce. In mathematical terms, alleles are subject to sampling error. As a result, when selective forces are absent or relatively weak, allele frequencies tend to "drift" upward or downward randomly (in a random walk). This drift halts when an allele eventually becomes fixed, either by disappearing from the population, or replacing the other alleles entirely. Genetic drift may therefore eliminate some alleles from a population due to chance alone. Even in the absence of selective forces, genetic drift can cause two separate populations which began with the same genetic structure to drift apart into two divergent populations with different sets of alleles.

The time for an allele to become fixed by genetic drift depends on population size, with fixation occurring more rapidly in smaller populations. The precise measure of populations that is important here is called the effective population size, which was defined by Sewall Wright as a theoretical number representing the number of breeding individuals that would exhibit the same observed degree of inbreeding.

Although natural selection is responsible for adaptation, the relative importance of the two forces of natural selection and genetic drift in driving evolutionary change in general is an area of current research in evolutionary biology. These investigations were prompted by the neutral theory of molecular evolution, which proposed that most evolutionary changes are the result of the fixation of neutral mutations that do not have any immediate effects on the fitness of an organism. Hence, in this model, most genetic changes in a population are the result of constant mutation pressure and genetic drift.

GENE FLOW

Male lions leave the pride where they are born and take over a new pride to mate. This results in gene flow between prides.Gene flow is the exchange of genes between populations, which are usually of the same species. Examples of gene flow within a species include the migration and then breeding of organisms, or the exchange of pollen. Gene transfer between species includes the formation of hybrid organisms and horizontal gene transfer.

Migration into or out of a population can change allele frequencies, as well as introducing genetic variation into a population. Immigration may add new genetic material to the established gene pool of a population. Conversely, emigration may remove genetic material. As barriers to reproduction between two diverging populations are required for the populations to become new species, gene flow may slow this process by spreading genetic differences between the populations. Gene flow is hindered by mountain ranges, oceans and deserts or even man-made structures such as the Great Wall of China, which has hindered the flow of plant genes.

Depending on how far two species have diverged since their most recent common ancestor, it may still be possible for them to produce offspring, as with horses and donkeys mating to produce mules. Such hybrids are generally infertile, due to the two different sets of chromosomes being unable to pair up during meiosis. In this case, closely related species may regularly interbreed, but hybrids will be selected against and the species will remain distinct. However, viable hybrids are occasionally formed and these new species can either have properties intermediate between their parent species, or possess a totally new phenotype. The importance of hybridization in creating new species of animals is unclear, although cases have been seen in many types of animals with the gray tree frog being a particularly well-studied example.

Hybridization is, however, an important means of speciation in plants, since polyploidy (having more than two copies of each chromosome) is tolerated in plants more readily than in animals. Polyploidy is important in hybrids as it allows reproduction, with the two different sets of chromosomes each being able to pair with an identical partner during meiosis. Polyploids also have more genetic diversity, which allows them to avoid inbreeding depression in small populations.

Horizontal gene transfer is the transfer of genetic material from one organism to another organism that is not its offspring; this is most common among bacteria. In medicine, this contributes to the spread of antibiotic resistance, as when one bacteria acquires resistance genes it can rapidly transfer them to other species. Horizontal transfer of genes from bacteria to eukaryotes such as the yeast. Saccharomyces cerevisiae and the adzuki bean beetle *Callosobruchus chinensis* may also have occurred. An example of larger-scale transfers are the eukaryotic bdelloid rotifers, which appear to have received a range of genes from bacteria, fungi, and plants. Viruses can also carry DNA between organisms, allowing transfer of genes even across biological domains. Large-scale gene transfer has also occurred between the ancestors of eukaryotic cells and prokaryotes, during the acquisition of chloroplasts and mitochondria.

Outcomes

Evolution influences every aspect of the form and behavior of organisms. Most prominent are the specific behavioral and physical adaptations that are the outcome of natural selection. These adaptations increase fitness by aiding activities such as finding food, avoiding predators or attracting mates. Organisms can also respond to selection by co-operating with each other, usually by aiding their relatives or engaging in mutually beneficial symbiosis. In the longer term, evolution produces new species through splitting ancestral populations of organisms into new groups that cannot or will not interbreed.

These outcomes of evolution are sometimes divided into macroevolution, which is evolution that occurs at or above the level of species, such as extinction and speciation, and microevolution, which is smaller evolutionary changes, such as adaptations, within a species or population. In general, macroevolution is regarded as the outcome of long periods of microevolution. Thus, the distinction between micro- and macroevolution is not a fundamental one - the difference is simply the time involved However, in macroevolution, the traits of the entire species may be important. For instance, a large amount of variation among individuals allows a species to rapidly adapt to new habitats, lessening the chance of it going extinct, while a wide geographic range increases the chance of speciation, by making it more likely that part of the population will become isolated. In this sense, microevolution and macroevolution might involve selection at different levels - with microevolution acting on genes and organisms, versus macroevolutionary processes acting on entire species and affecting the rate of speciation and extinction.

A common misconception is that evolution is "progressive," but natural selection has no long-term goal and does not necessarily produce greater complexity. Although complex species have evolved, this occurs as a side effect of the overall number of organisms increasing, and simple forms of life remain more common. For example, the overwhelming majority of species are microscopic prokaryotes, which form about half the world's biomass despite their small size and constitute the vast majority of Earth's biodiversity. Simple organisms have therefore been the dominant form of life on Earth throughout its history and continue to be the main form of life up to the present day, with complex life only appearing more diverse because it is more noticeable.

Adaptation

They are produced by a combination of the continuous production of small, random changes in traits, followed by natural selection of the variants best-suited for their environment. This process can cause either the gain of a new feature, or the loss of an ancestral feature. An example that shows both types of change is bacterial adaptation to antibiotic selection, with genetic changes causing antibiotic resistance by both modifying the target of the drug, or increasing the activity of transporters that pump the drug out of the cell. Other striking examples are the bacteria *Escherichia coli* evolving the ability to use citric acid as a nutrient in a long-term laboratory experiment or Flavobacterium evolving a novel enzyme that allows these bacteria to grow on the by-products of nylon manufacturing.

However, many traits that appear to be simple adaptations are in fact exaptations: structures originally adapted for one function, but which coincidentally became somewhat useful for some other function in the process One example is the African lizard *Holaspis guentheri*, which developed an extremely flat head for hiding in crevices, as can be seen by looking at its near relatives. However, in this species, the head has become so flattened that it assists in gliding from tree to tree—an exaptation. Another is the recruitment of enzymes from glycolysis and xenobiotic metabolism to serve as structural proteins called crystallins within the lenses of organisms' eyes.

As adaptation occurs through the gradual modification of existing structures, structures with similar internal organization may have very different functions in related organisms. This is the result of a single ancestral structure being adapted to function in different ways. The bones within bat wings, for example, are structurally similar to both human hands and seal flippers, due to the common descent of these structures from an ancestor that also had five digits at the end of each forelimb. Other idiosyncratic anatomical features, such as bones in the wrist of the panda being formed into a false "thumb", indicate that an organism's evolutionary lineage can limit what adaptations are possible.

During adaptation, some structures may lose their original function and become vestigial structures. Such structures may have little or no function in a current species, yet have a clear function in ancestral species, or other closely related species. Examples include pseudogenes, the non-functional remains of eyes in blind cave-dwelling fish wings in flightless birds and the presence of hip bones in whales and snakes. Examples of vestigial structures in humans include wisdom teeth, the coccyx and the vermiform appendix.

An area of current investigation in evolutionary developmental biology is the developmental basis of adaptations and exaptations. This research addresses the origin and evolution of embryonic development and how modifications of development and developmental processes produce novel features. These studies have shown that evolution can alter development to create new structures, such as embryonic bone structures that develop into the jaw in other animals instead forming part of the middle ear in mammals It is also possible for structures that have been lost in evolution to reappear due to changes in developmental genes, such as a mutation in chickens causing embryos to grow teeth similar to those of crocodiles. It is now becoming clear that most alterations in the form of organisms are due to changes in the level and timing of the expression of a small set of conserved genes.

Co-evolution

Interactions between organisms can produce both conflict and co-operation. When the interaction is between pairs of species, such as a pathogen and a host, or a predator and its prey, these species can develop matched sets of adaptations.

Here, the evolution of one species causes adaptations in a second species. These changes in the second species then, in turn, cause new adaptations in the first species. This cycle of selection and response is called co-evolution. An example is the production of tetrodotoxin in the rough-skinned newt and the evolution of tetrodotoxin resistance in its predator, the common garter snake. In this predator-prey pair, an evolutionary arms race has produced high levels of toxin in the newt and correspondingly high levels of resistance in the snake.

Co-operation

Many cases of mutually beneficial interactions have evolved. For instance, an extreme cooperation exists between plants and the mycorrhizal fungi that grow on their roots and aid the plant in absorbing nutrients from the soil. This is a reciprocal relationship as the plants provide the fungi with sugars from photosynthesis. Here, the fungi actually grow inside plant cells, allowing them to exchange nutrients with their hosts, while sending signals that suppress the plant immune system.

Coalitions between organisms of the same species have also evolved. An extreme case is the eusociality found in social insects, such as bees, termites and ants, where sterile insects feed and guard the small number of organisms in a colony that are able to reproduce. On an even smaller scale, the somatic cells that make up the body of an animal limit their reproduction so they can maintain a stable organism, which then supports a small number of the animal's germ cells to produce offspring. Here, somatic cells respond to specific signals that instruct them to either grow or kill themselves. If cells ignore these signals and attempt to multiply inappropriately, their uncontrolled growth causes cancer.

These examples of cooperation within species are thought to have evolved through the process of kin selection, which is where one organism acts to help raise a relative's offspring. This activity is selected for because if the helping individual contains alleles which promote the helping activity, it is likely that its kin will also contain these alleles and thus those alleles will be passed on. Other processes that may promote cooperation include group selection, where cooperation provides benefits to a group of organisms.

Speciation

Speciation is the process where a species diverges into two or more descendant species. It has been observed multiple times under both controlled laboratory conditions and in nature in sexually reproducing organisms, speciation results from reproductive isolation followed by genealogical divergence. There are four mechanisms for speciation. The most common in animals is allopatric speciation, which occurs in populations initially isolated geographically, such as by habitat fragmentation or migration. Selection under these conditions can produce very rapid changes in the appearance and behaviour of organisms As selection and drift act independently on populations isolated from the rest of their species, separation may eventually produce organisms that cannot interbreed.

The second mechanism of speciation is peripatric speciation, which occurs when small populations of organisms become isolated in a new environment. This differs from allopatric speciation in that the isolated populations are numerically much smaller than the parental population. Here, the founder effect causes rapid speciation through both rapid genetic drift and selection on a small gene pool.

The third mechanism of speciation is parapatric speciation. This is similar to peripatric speciation in that a small population enters a new habitat, but differs in that there is no physical separation between these two populations. Instead, speciation results from the evolution of mechanisms that reduce gene flow between the two populations Generally this occurs when there has been a drastic change in the environment within the parental species' habitat. One example is the grass *Anthoxanthum odoratum*, which can undergo parapatric speciation in response to localized metal pollution from mines Here, plants evolve that have resistance to high levels of metals in the soil. Selection against interbreeding with the metal-sensitive parental population produces a change in flowering time of the metal-resistant plants, causing reproductive isolation. Selection against hybrids between the two populations may cause reinforcement, which is the evolution of traits that promote mating within a species, as well as character displacement, which is when two species become more distinct in appearance.

Finally, in sympatric speciation species diverge without geographic isolation or changes in habitat. This form is rare since even a small amount of gene flow may remove genetic differences between parts of a population. Generally, sympatric speciation in animals requires the evolution of both genetic differences and non-random mating, to allow reproductive isolation to evolve.

One type of sympatric speciation involves cross-breeding of two related species to produce a new hybrid species. This is not common in animals as animal hybrids are usually sterile, because during meiosis the homologous chromosomes from each parent, being from different species cannot successfully pair. It is more common in plants, however because plants often double their number of chromosomes, to form polyploids. This allows the chromosomes from each parental species to form a matching pair during meiosis, as each parent's chromosomes is represented by a pair already An example of such a speciation event is when the plant species *Arabidopsis thaliana* and *Arabidopsis arenosa* cross-bred to give the new species *Arabidopsis suecica.* This happened about 20,000 years ago and the speciation process has been repeated in the laboratory, which allows the study of the genetic mechanisms involved in this process. Indeed, chromosome doubling within a species may be a common cause of reproductive isolation, as half the doubled chromosomes will be unmatched when breeding with undoubled organisms.

Speciation events are important in the theory of punctuated equilibrium, which accounts for the pattern in the fossil record of short "bursts" of evolution interspersed with relatively long periods of stasis, where species remain relatively unchanged. In this theory, speciation and rapid evolution are linked, with natural selection and genetic drift acting most strongly on organisms undergoing speciation in novel habitats or small populations. As a result, the periods of stasis in the fossil record correspond to the parental population, and the organisms undergoing speciation and rapid evolution are found in small populations or geographically restricted habitats, and therefore rarely being preserved as fossils.

Extinction

Extinction is the disappearance of an entire species. Extinction is not an unusual event, as species regularly appear through speciation, and disappear through extinction Indeed, virtually all animal and plant species that have lived on earth are now extinct, and extinction appears to be the ultimate fate of all species. These extinctions have happened continuously throughout the history of life, although the rate of extinction spikes in occasional mass extinction events. The Cretaceous–Tertiary extinction event, during which the dinosaurs went extinct, is the most well-known, but the earlier Permian–Triassic extinction event was even more severe, with approximately 96 per cent of species driven to extinction The Holocene extinction event is an ongoing mass extinction associated with humanity's expansion across the globe over the past few thousand years. Present-day extinction rates are 100-1000 times greater than the background rate, and up to 30 percent of species may be extinct by the mid 21st century Human activities are now the primary cause of the ongoing extinction event; global warming may further accelerate it in the future.

The role of extinction in evolution depends on which type is considered. The causes of the continuous "low-level" extinction events, which form the majority of extinctions, are not well understood and may be the result of competition between species for limited resources (competitive exclusion If competition from other species does alter the probability that a species will become extinct, this could produce species selection as a level of natural selection The intermittent mass extinctions are also important, but instead of acting as a selective force, they drastically reduce diversity in a nonspecific manner and promote bursts of rapid evolution and speciation in survivors.

Evolution as Theory and Fact

The potentially confusing statement that "evolution is both a theory and a fact" is frequently seen in biological literature. The point of this statement is to distinguish the two meanings of evolution. First, the "fact of evolution", the observed changes in populations of organisms over time. Second, a shortcut for the "theory of evolution", namely the modern evolutionary synthesis, which is the current scientific explanation for why these changes occur. Which meaning of evolution is intended, if it is not explicit, can be discerned by the context.

Evolution, Fact and Theory

Evolution has been described as "fact and theory," "fact not theory" and, "only a theory, not a fact". This illustrates a terminological confusion that hampers discussion The meanings of the terms "evolution" and "fact" and "theory", are described below:

Evolution

Evolution is usually defined simply as changes in trait or gene frequency in a population of organisms from one generation to the next. However, "evolution" is often used to include the following additional claims:

1. Differences in trait composition between isolated populations over many generations may result in the origin of new species.
2. All living organisms alive today have descended from a common ancestor (or ancestral gene pool).

According to Douglas Futuyma, 'biological evolution may be slight or substantial; it embraces everything from slight changes in the proportion of different alleles within a population (such as those determining blood types) to the successive alterations that led from the earliest proto-organism to snails, bees, giraffes, and dandelions.'

The term "evolution", especially when referred to as a "theory", is also used more broadly to incorporate processes such as natural selection and genetic drift.

Fact

Fact is often used by scientists to refer to experimental data or objective verifiable observations. "Fact" is also used in a wider sense to mean any hypothesis for which there is overwhelming evidence.

Evolution is a fact in the sense of it being overwhelmingly validated by the evidence. Frequently evolution is said to be a fact in the same way as the Earth revolving around the Sun is a fact The following quotation from H. J. Muller, *One Hundred Years Without Darwin Are Enough* explains the point.

There is no sharp line between speculation, hypothesis, theory, principle, and fact, but only a difference along a sliding scale, in the degree of probability of the idea. When we say a thing is a fact, then, we only mean that its probability is an extremely high one: so high that we are not bothered by doubt about it and are ready to act accordingly. Now in this use of the term fact, the only proper one, evolution is a fact.

The National Academy of Science (U.S.) makes a similar point:

> Scientists most often use the word "fact" to describe an observation. But scientists can also use fact to mean something that has been tested or observed so many times that there is no longer a compelling reason to keep testing or looking for examples. The occurrence of evolution in this sense is fact. Scientists no longer question whether descent with modification occurred because the evidence is so strong.
>
> Philosophers of science argue that we do not know anything with absolute certainty: even direct observations

may be "theory laden" and depend on assumptions about our senses and the measuring instruments used. In this sense all facts are provisional.

Theory

Scientific theories describe the coherent framework into which observable data fit. The scientific definition of the word "theory" is different from the colloquial sense of the word. Colloquially, "theory" can mean a conjecture, an opinion, or a speculation that does not have to be based on facts or make testable predictions. In science, the meaning of theory is more rigorous: a theory must be based on observed facts and make testable predictions.

In science, a current theory is a theory that has no equally acceptable or more acceptable alternative theory, and has survived attempts at falsification. That is, there have been no observations made which contradict it to this point and, indeed, every observation ever made either supports the current theory or at least does not falsify it by contradicting it completely. A revision of the current theory, or the generation of a new theory is necessary if new observations contradict the current theory, as the current findings are in need of a new explanation (see scientific revolution or paradigm shift). However, the falsification of a theory does not falsify the facts on which the theory is based.

Evolution Compared with Gravity

The terms "fact" and "theory" can be applied to evolution, just as they are to gravity Misuse and misunderstanding of how those terms are applied to evolution have been used to construct arguments disputing the validity of evolution.

There have been many theories that attempt to explain the fact of gravity. That is, scientists ask what gravity is, and what causes it. They develop a model to explain gravity, a theory of gravity. Many explanations of gravity that qualify as a *Theory of Gravity* have been proposed over the centuries: Aristotle's, Galileo's, Newton's, and now Einstein's. Confusion of the terms can arise when we use a single word to describe both the observed facts and the theory that explains it. The word "gravity" can be used to refer to the observed facts (i.e., the observed attraction of masses) and the theory used to explain it (gravity is the reason why masses attract each other). Thus, gravity is both a "theory" and a "fact."

In the study of biological species, the facts include fossils and measurements of these fossils. The location of a fossil is an example of a fact (using the scientific meaning of the word fact). In species that rapidly reproduce, for example fruit flies, the process of evolutionary change has been observed in the laboratory. The observation of fruit fly populations changing character is also an example of a fact. So evolution is a fact just as the observations of gravity are a fact.

In biology, there have been many attempts to explain these observations over the years. Lamarckism, Transmutationism and Orthogenesis were all non-Darwinian theories that attempted to explain the observations of species and fossils and other evidence. However, the Theory of Evolution is the explanation for all relevant observations regarding the development of life, based on a model that explains all the available data and observations. Thus, evolution is not only a fact but also a theory, just as gravity is both a fact and a theory.

Evolution as Theory and Fact in the Literature

The confusion between "fact" and "theory" and the use of the word "evolution" is largely due to some authors using evolution to refer to the changes that occur within species over generations and common descent, while others use the term more generally to include the mechanisms driving the change. However, among biologists at least, there seems to be consensus that evolution is a fact:

American zoologist and paleontologist George Simpson stated that

> 'Darwin . . . finally and definitely established evolution as a fact.'

H.J. Muller has written:

'If you like, then, I will grant you that in an absolute sense evolution is not a fact, or rather, that it is no more a fact than that you are hearing or reading these words.'

Kenneth R. Miller writes:

'evolution is as much a fact as anything we know in science.'

Ernst Mayr has observed:

'The basic theory of evolution has been confirmed so completely that most modern biologists consider evolution simply a fact. How else except by the word evolution can we designate the sequence of faunas and floras in precisely dated geological strata? And evolutionary change is also simply a fact owing to the changes in the content of gene pools from generation to generation'.

Evolution As Fact and Theory

Commonly "fact" is used to refer to the observable changes in organisms' traits over generations while the word "theory" is reserved for the mechanisms that cause these changes:

Paleontologist Stephen Jay Gould writes:

'Evolution is a theory. It is also a fact. And facts and theories are different things, not rungs in a hierarchy of increasing certainty. Facts are the world's data. Theories are structures of ideas that explain and interpret facts. Facts do not go away when scientists debate rival theories to explain them. Einstein's theory of gravitation replaced Newton's, but apples did not suspend themselves in mid-air, pending the outcome. And humans evolved from ape-like ancestors whether they did so by Darwin's proposed mechanism or by some other yet to be discovered.'

Similarly, biologist Richard Lenski says:

'Scientific understanding requires both facts and theories that can explain those facts in a coherent manner. Evolution, in this context, is both a fact and a theory. It is an incontrovertible fact that organisms have changed, or evolved, during the history of life on Earth. And biologists have identified and investigated mechanisms that can explain the major patterns of change.

Evolution as Fact not Theory

Other commentators, focusing on the changes in species over generations and in some cases common ancestry have stressed that evolution is a fact to emphasize the weight of supporting evidence while denying it is helpful to use the term "theory":

R.C. Lewontin wrote:

'It is time for students of the evolutionary process, especially those who have been misquoted and used by the creationists, to state clearly that evolution is a fact, not theory.'

Douglas Futuyma writes in his *Evolutionary Biology* book

'The statement that organisms have descended with modifications from common ancestors—the historical reality of evolution—is not a theory. It is a fact, as fully as the fact of the earth's revolution about the sun.'

Richard Dawkins says:

'One thing all real scientists agree upon is the fact of evolution itself. It is a fact that we are cousins of gorillas, kangaroos, starfish, and bacteria. Evolution is as much a fact as the heat of the sun. It is not a theory, and for pity's sake, let's stop confusing the philosophically naive by calling it so. Evolution is a fact.'

Neil Campbell wrote in his 1990 biology textbook:

'Today, nearly all biologists acknowledge that evolution is a fact. The term theory is no longer appropriate except when referring to the various models that attempt to explain how life evolves . . . it is important to understand that the current questions about how life evolves in no way implies any disagreement over the fact of evolution.'

Predictive Power

A central tenet in science is that a scientific theory is supposed to have predictive power, and verification of predictions are seen as an important and necessary support for the theory. The theory of evolution did provide such predictions. Three examples are:

Genetic information must be transmitted in a molecular way that will be almost exact but permit slight changes. Since this prediction was made, biologists have discovered the existence of DNA, which has a mutation rate of roughly 10-9 per nucleotide per cell division; this provides just such a mechanism.

Some DNA sequences are shared by very different organisms. It has been predicted by the theory of evolution that the differences in such DNA sequences between two organisms should roughly resemble both the biological difference between them according to their anatomy and the time that had passed since these two organisms have separated in the course of evolution, as seen in fossil evidence. The rate of accumulating such changes should be low for some sequences, which code for critical RNA or proteins, and high for others - that code for less critical RNA or proteins; but for every specific sequence, the rate of change should be roughly constant through evolution. These results have been experimentally confirmed. Two examples are DNA sequences coding for rRNA which is highly conserved, and DNA sequences coding for fibrinopeptides (amino acid chains which are discarded during the formation of fibrin), which are highly non-conserved.

Prior to 2004, paleontologists had found fossils of amphibians with necks, ears, and four legs, in rock no older than 365 million years old. In rocks more than 385 million years old they could only find fish, without these amphibian characteristics. Evolutionary theory predicted that an intermediate form between these two (i.e. a "fishibian") should be found in rock dated between 365 and 385 million years ago. In 2004, an expedition to islands in the Canadian arctic searching in rocks that were 375 million years old discovered fossils of Tiktaalik.

Chapter–4

Evidence of Common Descent

INTRODUCTION

The wide range of evidence of common descent of living things strongly indicates the occurrence of evolution and provides a wealth of information on the natural processes by which the variety of life on Earth developed, supporting the modern evolutionary synthesis.

Fossils are important for estimating when various lineages developed. As fossilization is an uncommon occurrence, usually requiring hard body parts and death near a site where sediments are being deposited, the fossil record only provides sparse and intermittent information about the evolution of life. Evidence of organisms prior to the development of hard body parts such as shells, bones and teeth is especially scarce, but exists in the form of ancient microfossils, as well as impressions of various soft-bodied organisms.

Comparison of the genetic sequence of organisms has revealed that organisms that are phylogenetically close have a higher degree of sequence similarity than organisms that are phylogenetically distant. Further evidence for common descent comes from genetic detritus such as pseudogenes, regions of DNA that are orthologous to a gene in a related organism, but are no longer active and appear to be undergoing a steady process of degeneration. Since metabolic processes do not leave fossils, research into the evolution of the basic cellular processes is done largely by comparison of existing organisms. Many lineages diverged at different stages of development, so it is possible to determine when certain metabolic processes appeared by comparing the traits of the descendants of a common ancestor.

Evidence from Genetics

Although it has only recently become available, the best evidence for common descent comes from the study of gene sequences. Comparative sequence analysis examines the relationship between the DNA sequences of different species, producing several lines of evidence that confirm Darwin's original hypothesis of common descent. If the hypothesis of common descent is true, then species that share a common ancestor will have inherited that ancestor's DNA sequence. Notably they will have inherited mutations unique to that ancestor. More closely-related species will have a greater fraction of identical sequence and will have shared substitutions when compared to more distantly-related species.

The simplest and most powerful evidence is provided by phylogenetic reconstruction. Such reconstructions, especially when done using slowly-evolving protein sequences, are often quite robust and can be used to reconstruct a great deal of the evolutionary history of modern organisms (and even in some instances such as the recovered gene sequences of mammoths, Neanderthals or T. rex, the evolutionary history of extinct organisms). These reconstructed phylogenies recapitulate the relationships established through morphological and biochemical studies. The most detailed reconstructions have been performed on the basis of the mitochondrial genomes shared by all eukaryotic organisms, which are short and easy to sequence; the broadest reconstructions have been performed either using the sequences of a few very ancient proteins or by using ribosomal RNA sequence.

This evidence does not support the rival hypothesis that genetic similarity of two species is the product of common functional or structural requirements, and not common descent (for example, if there is one best way to produce a hoof, all hoofed creatures will share a genetic basis even if they are not related). However, phylogenetic relationships also extend to a wide variety of non-functional sequence elements, including repeats, transposons, pseudogenes, and mutations in protein-coding sequences that do not result in changes in amino-acid sequence. While a minority of these elements might later be found to harbor function, in aggregate they demonstrate that identity must be the product of common descent rather than common function.

Finally, a deeper understanding of developmental biology shows that common morphology is, in fact, the product of shared genetic elements. For example, although camera-like eyes are believed to have evolved independently on many separate occasions, they share a common set of light-sensing proteins (opsins), suggesting a common point of origin for all sighted creatures. Another noteworthy example is the familiar vertebrate body plan, whose structure is controlled by the homeobox (Hox) family of genes.

EVIDENCE FROM PALEONTOLOGY

An insect trapped in amber.When organisms die, they often decompose rapidly or are consumed by scavengers, leaving no permanent evidences of their existence. However, occasionally, some organisms are preserved. The remains or traces of organisms from a past geologic age embedded in rocks by natural processes are called fossils. They are extremely important for understanding the evolutionary history of life on Earth, as they provide direct evidence of evolution and detailed information on the ancestry of organisms. Paleontology is the study of past life based on fossil records and their relations to different geologic time periods.

For fossilization to take place, the traces and remains of organisms must be quickly buried so that weathering and decomposition do not occur. Skeletal structures or other hard parts of the organisms are the most commonly occurring form of fossilized remains. There are also some trace "fossils" showing moulds, cast or imprints of some previous organisms.

As an animal dies, the organic materials gradually decay, such that the bones become porous. If the animal is subsequently buried in mud, mineral salts will infiltrate into the bones and gradually fill up the pores. The bones will harden into stones and be preserved as fossils. This process is known as petrification. If dead animals are covered by wind-blown sand, and if the sand is subsequently turned into mud by heavy rain or floods, the same process of mineral infiltration may occur. Apart from petrification, the dead bodies of organisms may be well preserved in ice, in hardened resin of coniferous trees (amber), in tar, or in anaerobic, acidic peat. Fossilization can sometimes be a trace, an impression of a form. Examples include leaves and footprints, the fossils of which are made in layers that then harden.

FOSSIL RECORDS

It is possible to find out how a particular group of organisms evolved by arranging its fossil records in a chronological sequence. Such a sequence can be determined because fossils are mainly found in sedimentary rock. Sedimentary rock is formed by layers of silt or mud on top of each other; thus, the resulting rock contains a series of horizontal layers, or strata. Each layer contains fossils which are typical for a specific time period during which they were made. The lowest strata contain the oldest rock and the earliest fossils, while the highest strata contain the youngest rock and more recent fossils.

A succession of animals and plants can also be seen from fossil records. By studying the number and complexity of different fossils at different stratigraphic levels, it has been shown that older fossil-bearing rocks contain fewer types of fossilized organisms, and they all have a simpler structure, whereas younger rocks contain a greater variety of fossils, often with increasingly complex structures.

In the past, geologists could only roughly estimate the ages of various strata and the fossils found. They did so, for instance, by estimating the time for the formation of sedimentary rock layer by layer. Today, by measuring the proportions of radioactive and stable elements in a given rock, the ages of fossils can be more precisely dated by scientists. This technique is known as radiometric dating.

Throughout the fossil record, many species that appear at an early stratigraphic level disappear at a later level. This is interpreted in evolutionary terms as indicating the times at which species originated and became extinct. Geographical regions and climatic conditions have varied throughout the Earth's history. Since organisms are adapted to particular environments, the constantly changing conditions favoured species which adapted to new environments through the mechanism of natural selection.

According to fossil records, some modern species of plants and animals are found to be almost identical to the species that lived in ancient geological ages. They are existing species of ancient lineages that have remained morphologically (and probably also physiologically) somewhat unchanged for a very long time. Consequently, they are called "living fossils" by laypeople. Examples of "living fossils" include the tuatara, the nautilus, the horseshoe crab, the coelacanth, the ginkgo, the Wollemi pine, and the metasequoia.

EXTENT OF THE FOSSIL RECORD

Cynognathus, a Eucynodont, one of a grouping of Therapsids ("mammal-like reptiles") that is ancestral to all modern mammals.Despite the relative rarity of suitable conditions for fossilization, approximately 250,000 fossil species are known. The number of individual fossils this represents varies greatly from species to species, but many millions of fossils have been recovered: for instance, more than three million fossils from the last Ice Age have been recovered from the La Brea Tar Pits in Los Angeles Many more fossils are still in the ground, in various geological formations known to contain a high fossil density, allowing estimates of the total fossil content of the formation to be made. An example of this occurs in South Africa's Beaufort Formation (part of the Karoo Supergroup, which covers most of South Africa), which is rich in vertebrate fossils, including therapsids (reptile/mammal transitional forms). It has been estimated that this formation contains 800 billion vertebrate fossils.

While the fossils cannot undoubtedly prove common descent, they are highly suggestive of it if they can provide two things: (1) Older forms are simpler than newer forms; (2) If there is a gradual sequence from a few ancient species to a more vast number of modern ones. The fossil record certainly meets the first criterion. Among the earliest mammalian fossils, there are none that indicate the presence of specialized mammals like whales, but we do find fossils of whale-like terrestrial mammals that possessed underdeveloped legs. The second criterion poses a sort of impasse between evolutionary scientists who claim their findings to be incomplete yet compelling and creationists who bemoan them as severly lacking.

In fact, it is this scarceness within the record that led paleontologists Stephen Jay Gould and Niles Eldredge to propose their theory of punctuated equilibrium. Gould writes that certain features of the fossil record are, in fact, inconsistent with gradualism, the more mainstream approach to evolution. He insists that because certain species show no morphological changes throughout their existance (from origination to extinction) and the apparant "sudden appearance" of certain species, rather than a gradual transformation, it must be that evolution occurs too quickly to leave a trace in the fossil record. He, along with Eldredge, therefore proposed that evolution occurs very quickly over short (i.e. a few thousand years) periods of time.

Gould and Eldredge's claims are only that the minimal availability of fossils of transitional forms points to periods of very rapid evolution interrupted by much longer periods of preservation of form; they do not, however, argue against common descent. It is, therefore, extremely unfortunate that many anti-evolutionists misuse the theory of punctuated equilibrium to bolster their attacks on common ancestry in specific and evolution in general. This is because "the scarcity [of transitional fossils] has . . . implications for certain theories of evolutionary mechanisms and and rates of evolution.

Evolution of the Horse

Evolution of the horse showing reconstruction of the fossil species obtained from successive rock strata. The foot diagrams are all front views of the left forefoot. The third metacarpal is shaded throughout. The teeth are shown in longitudinal section. Due to an almost-complete fossil record found in North American sedimentary deposits from the early Eocene to the present, the horse provides one of the best examples of evolutionary history (phylogeny).

This evolutionary sequence starts with a small animal called *Hyracotherium* (commonly referred to as Eohippus) which lived in North America about 54 million years ago, then spread across to Europe and Asia. Fossil remains of Hyracotherium show it to have differed from the modern horse in three important respects: it was a small animal (the size of a fox), lightly built and adapted for running; the limbs were short and slender, and the feet elongated so that the digits were almost vertical, with four digits in the forelimbs and three digits in the hindlimbs; and the incisors were small, the molars having low crowns with rounded cusps covered in enamel.

The probable course of development of horses from Hyracotherium to Equus (the modern horse) involved at least 12 genera and several hundred species.

Fossilized plants found in different strata show that the marshy, wooded country in which Hyracotherium lived became gradually drier. Survival now depended on the head being in an elevated position for gaining a good view of the surrounding countryside, and on a high turn of speed for escape from predators, hence the increase in size and the replacement of the splayed-out foot by the hoofed foot. The drier, harder ground would make the original splayed-out foot unnecessary for support. The changes in the teeth can be explained by assuming that the diet changed from soft vegetation to grass. A dominant genus from each geological period has been selected to show the progressive development of the horse.

Limitations

The fossil record is an important source for scientists when tracing the evolutionary history of organisms. However, because of limitations inherent in the record, there are not fine scales of intermediate forms between related groups of species. This lack of continuous fossils in the record is a major limitation in tracing the descent of biological groups. Furthermore, there are also much larger gaps between major evolutionary lineages. When transitional fossils are found that show intermediate forms in what had previously been a gap in knowledge, they are often popularly referred to as "missing links".

There is a gap of about 100 million years between the early Cambrian period and the later Ordovician period. The early Cambrian period was the period from which numerous fossils of sponges, cnidarians (e.g., jellyfish), echinoderms (e.g., eocrinoids), molluscs (e.g., snails) and arthropods (e.g., trilobites) are found. In the later Ordovician period, the first animal that really possessed the typical features of vertebrates, the Australian fish, Arandaspis appeared. Thus few, if any, fossils of an intermediate type between invertebrates and vertebrates have been found, although likely candidates include the Burgess Shale animal, *Pikaia gracilens*, and its Maotianshan shales relatives, Myllokunmingia, Yunnanozoon, Haikouella lanceolata, and Haikouichthys.

In general, the probability that an organism becomes fossilized after death is very low;

Some species or groups are less likely to become fossils because they are soft-bodied;

Some species or groups are less likely to become fossils because they live (and die) in conditions that are not favourable for fossilization to occur in;

Many fossils have been destroyed through erosion and tectonic movements;

Some fossil remains are complete, but most are fragmentary;

Some evolutionary change occurs in populations at the limits of a species' ecological range, and as these populations are likely to be small, the probability of fossilization is lower.

Similarly, when environmental conditions change, the population of a species is likely to be greatly reduced, such that any evolutionary change induced by these new conditions is less likely to be fossilized. Most fossils convey information about external form, but little about how the organism functioned. Using present-day biodiversity as a guide, this suggests that the fossils unearthed represent only a small fraction of the large number of species of organisms that lived in the past.

Evidence from Comparative Anatomy

Comparative study of the anatomy of groups of animals or plants reveals that certain structural features are basically similar. For example, the basic structure of all flowers consists of sepals, petals, stigma, style and ovary; yet the size, colour, number of parts and specific structure are different for each individual species.

If widely separated groups of organisms are originated from a common ancestry, they are expected to have certain basic features in common. The degree of resemblance between two organisms should indicate how closely related they are in evolution.

Groups with little in common are assumed to have diverged from a common ancestor much earlier in geological history than groups which have a lot in common.

In deciding how closely related two animals are, a comparative anatomist looks for structures that are fundamentally similar, even though they may serve different functions in the adult. Such structures are described as homologous and suggest a common origin.

In cases where the similar structures serve different functions in adults, it may be necessary to trace their origin and embryonic development. A similar developmental origin suggests they are the same structure, and thus likely to be derived from a common ancestor.

When a group of organisms share a homologous structure which is specialized to perform a variety of functions in order to adapt different environmental conditions and modes of life are called adaptive radiation. The gradual spreading of organisms with adaptive radiation is known as divergent evolution.

Pentadactyl Limb

The pattern of limb bones called pentadactyl limb is an example of homologous structures. It is found in all classes of tetrapods (i.e. from amphibians to mammals). It can even be traced back to the fins of certain fossil fishes from which the first amphibians are thought to have evolved. The limb has a single proximal bone (humerus), two distal bones (radius and ulna), a series of carpals (wrist bones), followed by five series of metacarpals (palm bones) and phalanges (digits). Throughout the tetrapods, the fundamental structures of pentadactyl limbs are the same, indicating that they originated from a common ancestor. But in the course of evolution, these fundamental structures have been modified. They have become superficially different and unrelated structures to serve different functions in adaptation to different environments and modes of life. This phenomenon is clearly shown in the forelimbs of mammals. For example:

> In the monkey, the forelimbs are much elongated to form a grasping hand for climbing and swinging among trees.
>
> In the pig, the first digit is lost, and the second and fifth digits are reduced. The remaining two digits are longer and stouter than the rest and bear a hoof for supporting the body.

In the horse, the forelimbs are adapted for support and running by great elongation of the third digit bearing a hoof. The mole has a pair of short, spade-like forelimbs for burrowing.

The anteater uses its enlarged third digit for tearing down ant hills and termite nests. In the whale, the forelimbs become flippers for steering and maintaining equilibrium during swimming.

In the bat, the forelimbs have turned into wings for flying by great elongation of four digits, while the hook-like first digit remains free for hanging from trees.

Other Arthropod Appendages

Insect mouthparts and antennae are considered homologues of insect legs. Parallel developments are seen in some arachnids: The anterior pair of legs may be modified as analogues of antennae, particularly in whip scorpions, which walk on six legs. These developments provide support for the theory that complex modifications often arise by duplication of components, with the duplicates modified in different directions.

Under similar environmental conditions, fundamentally different structures in different groups of organisms may undergo modifications to serve similar functions. This phenomenon is called convergent evolution. Similar structures, physiological processes or mode of life in organisms apparently bearing no close phylogenetic links but showing adaptations to perform the same functions are described as analogous, for example:

Wings of bats, birds and insects;

the jointed legs of insects and vertebrates;

tail fin of fish, whale and lobster;

eyes of the vertebrates and cephalopod molluscs (squid and octopus). The difference between an inverted and non-inverted retina, the sensory cells lying beneath the nerve fibres. This results in the sensory cells being absent where the optic nerve is attached to the eye, thus creating a blind spot. The octopus eye has a non-inverted retina in which the sensory cells lie above the nerve fibres. There is therefore no blind spot in this kind of eye. Apart from this difference the two eyes are remarkably similar, an example of convergent evolution.

The strongest direct evidence for common descent comes from vestigial structures and embryonic development. Rudimentary body parts, those that are smaller and simpler in structure than corresponding parts in the ancestral species, are called vestigial organs. They are usually degenerated or underdeveloped. The existence of vestigial organs can be explained in terms of changes in the environment or modes of life of the species. Those organs are thought to be functional in the ancestral species but have now become unnecessary and non-functional. Examples are the vestigial hind limbs and pelvic girdles of whales, the haltere (vestigial hind wings) of flies and mosquitos, vestigial wings of flightless birds such as ostriches, the extra toes of ungulates that do not even reach the ground and the vestigial leaves of some xerophytes (e.g. cactus) and parasitic plants (e.g. dodder). It must be noted, however, that vestigial structures have lost the original function but may have another one. For example the halteres in dipterists help balance the insect while in flight and the wings of ostriches are used in mating rituals.

The most reasonable conclusion to draw is that these creatures descended from creatures in which these parts were functional, which in turn indicates that most (or indeed all) creatures descended from common ancenstors.

Evidence from Geographical Distribution

Data about the presence or absence of species on various continents and islands (biogeography) can provide evidence of common descent and shed light on patterns of speciation.

Continental Distribution

All organisms are adapted to their environment to a greater or lesser extent. If the abiotic and biotic factors within a habitat are capable of supporting a particular species in one geographic area, then one might assume that the same species would be found in a similar habitat in a similar geographic area, e.g. in Africa and South America. This is not the case. Plant and animal species are discontinuously distributed throughout the world:

Even greater differences can be found if Australia is taken into consideration, though it occupies the same latitude as much of South America and Africa. Marsupials like the kangaroo, the wallaby, and the wombat make up over 80 percent of Australia's indigenous mammal population. By contrast, marsupials are totally absent from Africa and are only represented by the opossum in South America and the Virginia Opossum in North America: The echidna and platypus, the only living representatives of primitive egg-laying mammals (monotremes), can be found only in Australia and are totally absent in the rest of the world. On the other hand, Australia has very few placental mammals and most of these either migrated from elsewhere (e.g. bats) or were introduced by human beings (e.g. rabbits).

South America via the land bridge in the Bering Strait and Isthmus of Panama; A large number of families of South American marsupials became extinct as a result of competition with these North American counterparts.

To Africa via the Strait of Gibraltar; and to Australia via South East Asia to which it was at one time connected by land The shallowness of the Bering Strait would have made the passage of animals between two northern continents a relatively easy matter, and it explains the present-day similarity of the two faunas. But once they had got down into the southern continents, they presumably became isolated from each other by various types of barriers. The submersion of the Isthmus of Panama: isolates the South American fauna. The Mediterranean Sea and the North African desert: partially isolate the African fauna. The submersion of the original connection between Australia and South East Asia: isolates the Australian fauna. Once isolated, the animals in each continent have shown adaptive radiatio to evolve along their own lines.

Chapter–5

HERITABILITY

INTRODUCTION

In genetics, Heritability is the proportion of phenotypic variation in a population that is attributable to genetic variation among individuals. Variation among individuals may be due to genetic and/or environmental factors. Heritability analyses estimate the relative contributions of differences in genetic and non-genetic factors to the total phenotypic variance in a population.

Definition

Relationship of phenotypic values to additive and dominance effects using a completely dominant locus.Consider a statistical model for describing some particular phenotype

Phenotype (P) = Genotype (G) + Environment (E).

Considering variances (Var), this becomes:

Var(P) = Var(G) + Var(E) + 2 Cov(G, E).

In planned experiments, we can often take Cov (G, E) = 0. Heritability is then defined as:

$$H^2 = \frac{Var(G)}{Var(P)}$$

The parameter H^2 is the broad-sense heritability and reflects all possible genetic contributions to a population's phenotypic variance. Included are effects due to allelic variation (additive variance), dominance variation or which act epistatically (multi-genic interactions), as well as maternal and paternal effects, where individuals are directly affected of their parents' phenotype (such as with milk production in mammals).

These additional terms can be included in genetic models. For example, the simplest genetic model involves a single locus with two alleles that effect some quantitative phenotype. We can calculate the linear regression of phenotype on the number of B alleles (0, 1, or 2), which is shown as the Linear Effect line. For any genotype, B_iB_j, the expected phenotype can then be written as the sum of the overall mean, a linear effect, and a dominance deviation:

$P_{ij} = \mu + a_i + a_j + d_{ij}$ = Population mean + Additive Effect ($a_{ij} = a_i + a_j$) + Dominance Deviation (d_{ij}).

The additive genetic variance is the weighted average of the squares of the additive effects :

$$Var(A) = f(bb)a_{bb}^2 + f(Bb)a_{Bb}^2 + f(BB)a_{BB}^2,$$

where f(bb)abb + f(Bb)aBb + f(BB)aBB = 0.

There is a similar relationship for variance of dominance deviations:

$$Var(D) = f(bb)d_{bb}^2 + f(Bb)d_{Bb}^2 + f(BB)d_{BB}^2,$$

where f(bb)dbb + f(Bb)dBb + f(BB)dBB = 0.

Narrow-sense heritability is defined as

$$h^2 = \frac{Var(A)}{Var(P)}$$

and quantifies only the portion of the phenotypic variation that is additive (allelic) by nature (note upper case H^2 for broad sense, lower case h^2 for narrow sense). When interested in improving livestock via artificial selection, for example, knowing the narrow-sense heritability of the trait of interest will allow predicting how much the mean of the trait will increase in the next generation as a function of how much the mean of the selected parents differs from the mean of the population from which the selected parents were chosen. The observed response to selection leads to an estimate of the narrow-sense heritability (called realized heritability).

ESTIMATING HERITABILITY

Estimating heritability is not a simple process, since only P can be observed or measured directly. Measuring the genetic and environmental variance requires various sophisticated statistical methods. These methods give better estimates when using data from closely related individuals - such as brothers, sisters, parents and offspring, rather than from more distantly related ones. The standard error for heritability estimates are generally very poor unless the dataset is large.

In non-human populations it is often possible to collect information in a controlled way. For example, among farm animals it is easy to arrange for a bull to produce offspring from a large number of cows. Due to ethical concerns, such a degree of experimental control is impossible when gathering human data. As a result, studies of human heritability often contrast identical twins who have been separated early in life and raised in different environments Such individuals have identical genotypes and can be used to separate the effects of genotype and environment. Twin studies entail problems of its own, such as: independently raised twins shared a common prenatal environment; they may have undergone intrauterine competition; the mother may be more physically stressed (less nutrients); and twins reared apart are difficult to find, and may reflect certain types of environments.

Heritability estimates are always relative to the genetic and environmental factors in the population, and are not absolute measurements of the contribution of genetic and environmental factors to a phenotype. Heritability estimates reflect the amount of variation in genotypic effects compared to variation in environmental effects. Heritability can be made larger by diversifying the genetic background, e.g., by using only very outbred individuals (which increases the Variance(G)) and/or by minimizing environmental effects (which decreases the Variance(E)). Smaller heritability, on the other hand, can be generated by using inbred individuals (which decreases the Variance(G)) or individuals reared in very diverse environments (which increases the Variance(E)). Due to such effects, different populations of a species might have different heritabilities even for the same trait.

In observational studies G and E may be correlated, giving rise to gene environment correlation. Depending on the methods used to estimate heritability, correlations between genetic factors and shared or non-shared environments may or may not be included in the total heritability estimate.

Because of the contextual nature of measured heritabilities, paradoxes often arise. For example, the heritability of a trait could be near 100% in one study and close to zero in another. In one study, e.g., a group of unrelated army recruits may be given identical training and nutrition and then their muscular strength may be measured. The variation in strength observed after the (identical) training will translate into a high heritability estimate. In another study, whose purpose might be to assess the efficacy of various workout regimes or nutritional programs, study subjects may be first chosen to match each other as closely as possible in prior physical characteristics before some of them are put onto Program A and others onto Program B, and this will lead to a low heritability estimate.

Heritability estimates are often misinterpreted. Heritability refers to the proportion of variation between individuals in a population that is influenced by genetic factors. Heritability describes the population, not individuals within that population. For example, It is incorrect to say that since the heritability of a personality trait is about .6, that means that 60% of your personality is inherited from your parents and 40% comes from the environment. The heritability estimate changes according to the genetic and environmental variability present in the population. In studies of genetically identical inbred animals, all traits have zero heritability. Heritability estimates can be much higher in outbred (genetically variable) populations under very homogeneous environments.

A highly genetically loaded trait (such as eye color) still assumes environmental input within normal limits (a certain range of temperature, oxygen in the atmosphere, etc.). A more useful distinction than "nature vs. nurture" is "obligate vs. facultative" — under typical environmental ranges, what traits are more "obligate" (e.g., the nose — everyone has a nose) or more "facultative" (sensitive to environmental variations, such as specific language learned during infancy). Another useful distinction is between traits that are likely to be adaptations (such as the nose) vs. those that are byproducts of adaptations (such the white color of bones), or are due to random variation (non-adaptive variation in, say, nose shape or size).

ESTIMATION METHODS

There are essentially two schools of thought regarding estimation of heritability.

One school of thought was developed by Sewall Wright at The University of Chicago, and further popularized by C.C. Li (University of Chicago) and J.L. Lush (Iowa State University). It is based on the analysis of correlations and, by extension, regression. Path Analysis was developed by Sewall Wright as a way of estimating heritability.

The second was originally developed by R. A. Fisher and expanded at The University of Edinburgh, Iowa State University, and North Carolina State University, as well as other schools. It is based on the analysis of variance of breeding studies, using the intraclass correlation of relatives. Various methods of estimating components of variance (and, hence, heritability) from ANOVA are used in these analyses.

Regression/correlation Methods of Estimation

The first school of estimation uses regression and correlation to estimate heritability.

Selection Experiments

Strength of selection (S) and response to selection (R) in an artificial selection experiment, h^2=R/S.Calculating the strength of selection, S (the difference in mean trait between the population as a whole and the selected parents of the next generation, also called the selection differential) and response to selection R (the difference in offspring and whole parental generation mean trait) in an artificial selection experiment will allow calculation of realized heritability as the response to selection relative to the strength of selection, h^2=R/S.

Comparison of Close relatives

In the comparison of relatives, we find that in general

$$h^2 = \frac{b}{r} = \frac{t}{r}$$

where r can be thought of as the coefficient of relatedness, b is the coefficient of regression and t the coefficient of correlation.

Parent-offspring Regression

Sir Francis Galton's (1889) data showing the relationship between offspring height (928 individuals) as a function of mean parent height (205 sets of parents). Heritability may be estimated by comparing parent and offspring traits. The slope of the line (0.57) approximates the heritability of the trait when offspring values are regressed against the average trait in the parents. If only one parent's value is used then heritability is twice the slope. (note that this is the source of the term "regression", since the offspring values always tend to regress to the mean value for the population, i.e., the slope is always less than one).

Full-sib Comparison

Full-sib designs compare phenotypic traits of siblings that share a mother and a father with other sibling groups. The estimate of the sibling phenotypic correlation is an index on familiality which is equal to half the additive genetic variance plus the common environment variance when there is only additive gene action.

Twin Studies

Heritability for traits in humans is most frequently estimated by comparing resemblances between twins. Identical twins (MZ twins) are twice as genetically similar as fraternal twins (DZ twins) and so heritability is approximately twice the difference in correlation between MZ and DZ twins, $h^2 = 2(r(MZ)-r(DZ))$. The effect of shared environment, c^2, contributes to similarity between siblings due to the commonality of the environment they are raised in. Shared environment is approximated by the DZ correlation minus half heritability, which is the degee to which DZ twins share the same genes, $c^2 = DZ-1/2h^2$. Unique environmental variance, e^2, reflects the degree to which identical twins raised together are dissimilar, $e^2 = 1-r(MZ)$.

The methodology of the classical twin study has been criticized, but these criticisms do not take into account the methodological innovations and refinements described above.

Analysis of Variance Methods of Estimation

The second set of methods of estimation of heritability involves ANOVA and estimation of variance components.

Basic Model

We use the basic discussion of Kempthorne (1957 [1969]). Considering only the most basic of genetic models, we can look at the quantitative contribution of a single locus with genotype Gi as

$$y_i = \mu + g_i + e$$

where

g_i is the effect of genotype G_i and e is the environmental effect.

Consider an experiment with a group of sires and their progeny from random dams. Since the progeny get half of their genes from the father and half from their (random) mother, the progeny equation.

INTRACLASS CORRELATIONS

Consider the experiment above. We have two groups of progeny we can compare. The first is comparing the various progeny for an individual sire (called within sire group). The variance will include terms for genetic variance (since they did not all get the same genotype) and environmental variance. This is thought of as an error term.

The second group of progeny are comparisons of means of half sibs with each other (called among sire group). In addition to the error term as in the within sire groups, we have an addition term due to the differences among different means of half sibs. The intraclass correlation is

$$corr(z, z^{I}) = corr(\mu + \frac{1}{2}g + e, \mu + \frac{1}{2}g + e') = \frac{1}{4}Vg$$

Model with Additive and Dominance Terms

For a model with additive and dominance terms, but not others, the equation for a single locus is

$y_{ij} = \mu + a_i + a_j + d_{ij} + e,$

where

ai is the additive effect of the ith allele, a_j is the additive effect of the j_{th} allele, dij is the dominance deviation for the ijth genotype, and e is the environment.

Using different relationship groups, we can evaluate different intraclass correlations. Using V_a as the additive genetic variance and V_d as the dominance deviation variance, intraclass correlations become linear functions of these parameters. In general;

Intraclass correlation = $rV_a + ?V_d$,

where r and ? are found as

r = P [alleles drawn at random from the relationship pair are identical by descent], and

? = P [genotypes drawn at random from the relationship pair are identical by descent].

Larger Models

When a large, complex pedigree is available for estimating heritability, the most efficient use of the data is in a restricted maximum likelihood (REML) model. The raw data will usually have three or more datapoints for each individual: a code for the sire, a code for the dam and one or several trait values. Different trait values may be for different traits or for different timepoints of measurement. The currently popular methodology relies on high degrees of certainty over the identities of the sire and dam; it is not common to treat the sire identity probabilistically. This is not usually a problem, since the methodology is rarely applied to wild populations (although it has been used for several wild ungulate and bird populations), and sires are invariably known with a very high degree of certainty in breeding programmes. There are also algorithms that account for uncertain paternity.

The pedigrees can be viewed using programs such as Pedigree Viewer and analysed with programs such as ASReml, VCE WOMBAT or BLUPF90 family's programs

Response to Selection

In selective breeding of plants and animals, the expected response to selection can be estimated by the following equation:

$$R = h^2S$$

In this equation, the Response to Selection (R) is defined as the realized average difference between the parent generation and the next generation. The Selection Differential (S) is defined as the average difference between the parent generation and the selected parents.

For example, imagine that a plant breeder is involved in a selective breeding project with the aim of increasing the number of kernels per ear of corn. For the sake of argument, let us assume that the average ear of corn in the parent generation has 100 kernels. Let us also assume that the selected parents produce corn with an average of 120 kernels per ear. If h^2 equals 0.5, then the next generation will produce corn with an average of 0.5(120-100) = 10 additional kernels per ear. Therefore, the total number of kernels per ear of corn will equal, on average, 110.

The concept of heritability plays a central role in the psychology of individual differences. Heritability has two definitions. The first is a statistical definition, and it defines heritability as the proportion of phenotypic variance attributable to genetic variance. The second definition is more common "sensical". It defines heritability as the extent to which genetic individual differences contribute to individual differences in observed behavior (or phenotypic individual differences). You should memorize both of these definitions.

Because heritability is a proportion, its numerical value will range from 0.0 (genes do not contribute at all to phenotypic individual differences) to 1.0 (genes are the only reason for individual differences). For human behavior, almost all estimates of heritability are in the moderate range of .30 to .60.

The quantity (1.0 - heritability) gives the environmentability of the trait. Environmentability has an analogous interpretation to heritability. It is the proportion of phenotypic variance attributable to environmental variance or the extent to which individual differences in the environment contribute to individual differences in behavior. If the heritability of most human behaviors is in the range of .30 to .60, then the environmentability of most human behaviors will be in the range of .40 to .70.

There are five important attributes about estimates of heritability and environmentability. They are:

1. Heritability and environmentability are abstract concepts. No matter what the numbers are, heritability estimates tell us nothing about the specific genes that contribute to a trait. Similarly, a numerical estimate of environmentability provides no information about the important environmental variables that influence a behaviour.

2. Heritability and environmentability are population concepts. They tell us nothing about an individual.

A heritability of .40 informs us that, on average, about 40% of the individual differences that we observe in, say, shyness may in some way be attributable to genetic individual difference. It does NOT mean that 40% of any person's shyness is due to his/her genes and the other 60% is due to his/her environment.

3. Heritability depends on the range of typical environments in the population that is studied. If the environment of the population is fairly uniform, then heritability may be high, but if the range of environmental differences is very large, then heritability may be low. In different words, if everyone is treated the same environmentally, then any differences that we observe will largely be due to genes; heritability will be large in this case. However, if the environment treats people very differently, then heritability may be small.

4. Environmentability depends on the range of genotypes in the population studied. This is the converse of the point made above. However, it probably does not apply strongly to human behavior as it does to the behavior of specially bred animals. Few—if any—human populations are as genetically homogeneous as breeds of dogs, sheep, etc.

5. Heritability is no cause for therapeutic nihilism. Because heritability depends on the range of typical environments in the population studied, it tells us little about the extreme environmental interventions utilized in some therapies.

PHENOTYPE

A phenotype is any observable characteristic or trait of an organism: such as its morphology, development, biochemical or physiological properties, or behaviour. Phenotypes result from the expression of an organism's genes as well as the influence of environmental factors and possible interactions between the two. The genotype of an organism are the inherited instructions it carries within its genetic code. Not all organisms with the same genotype look or act the same way, because appearance and behavior are modified by environmental and developmental conditions. Also in the same way, not all organisms that look alike necessarily have the same genotype. This genotype-phenotype distinction was proposed by Wilhelm Johannsen in 1911 to make clear the difference between an organism's heredity and what that heredity produces. The distinction is similar to that proposed by August Weismann, who distinguished between germ plasm (heredity) and somatic cells (the body). A more modern version is Francis Crick's Central dogma of molecular biology.

Despite its seemingly straightforward definition, the concept of the phenotype has some hidden subtleties. First, most of the molecules and structures coded by the genetic material are not visible in the appearance of an organism, yet they are observable (for example by Western blotting) and are thus part of the phenotype. Human blood groups are an example. So, by extension, the term phenotype must include characteristics that can be made visible by some technical procedure. Another extension adds behaviour to the phenotype since behaviours are also affected by both genotypic and environmental factors.

Second, the phenotype is not simply a product of the genotype, but is influenced by the environment to a greater or lesser extent. And, further, if the genotype is defined narrowly, then it must be remembered that not all heredity is carried by the nucleus. For example, mitochondria transmit their own DNA directly, not via the nucleus, though they divide in unison with the nucleus.

The phenotype is composed of traits or characteristics. Some phenotypes are controlled entirely by the individual's genes. Others are controlled by genes but are significantly affected by extragenetic or environmental factors. Almost all humans inherit the capacity to speak and understand language, but which language they learn is entirely an environmental matter.

Phenotypic Variation

Phenotypic variation (due to underlying heritable genetic variation) is a fundamental prerequisite for evolution by natural selection. It is the living organism as a whole that contributes (or not) to the next generation, so natural selection affects the genetic structure of a population indirectly via the contribution of phenotypes. Without phenotypic variation, there would be no evolution by natural selection.

The interaction between genotype and phenotype has often been conceptualized by the following relationship:

genotype + environment ? phenotype

A slightly more nuanced version of the relationships is:

genotype + environment + random-variation ? phenotype

Genotypes often have great flexibility in the modification and expression of phenotypes, in many organisms these phenotypes are very different under varying environmental conditions. The plant *Hieracium umbellatum* is found growing in two different habitats in Sweden. One habitat is rocky, sea-side cliffs, where the plants are bushy with broad leaves and expanded inflorescences; the other is among sand dunes where the plants grow prostrate with narrow leaves and compact inflorescences. These habitats alternate along the coast of Sweden and the habitat that the seeds of Hieracium umbellatum land in, determine the phenotype that grows.

An example of random variation in Drosophila flies is the number of ommatidia, which may vary (randomly) between left and right eyes in a single individual as much as they do between different genotypes overall, or between clones raised in different environments.

A phenotype is any detectable characteristic of an organism (i.e., structural, biochemical, physiological, and behavioral) determined by an interaction between its genotype and environment (of this distinction).

According to the autopoietic notion of living systems by Humberto Maturana, the phenotype is epigenetically being constructed throughout ontogeny, and we as observers make the distinctions that define any particular trait at any particular state of the organism's life cycle.

The idea of the phenotype has been generalized by Richard Dawkins in The Extended Phenotype to mean all the effects a gene has on the outside world that may influence its chances of being replicated. These can be effects on the organism in which the gene resides, the environment, or other organisms. For instance, a beaver dam might be considered a phenotype of beaver genes, the same way beaver's powerful incisor teeth are phenotype expressions of their genes. Dawkins also cites the effect of an organism on the behaviour of another organism, such as the devoted nurturing of a cuckoo by a parent clearly of a different species as an example of the extended phenotype.

The concept of phenotype can be extended to variations below the level of the gene that affect an organism's fitness. For example, silent mutations that do not change the corresponding amino acid sequence of a gene may change the frequency of guanine-cytosine base pairs (GC content). These base pairs have a higher thermal stability (melting point, see also DNA-DNA hybridization) than adenine-thymine, a property that might convey, among organisms living in high-temperature environments, a selective advantage on variants enriched in GC content.

GENOTYPE

The genotype is the genetic constitution of a cell, an organism, or an individual (i.e. the specific allele makeup of the individual) usually with reference to a specific character under consideration. For instance, the human albino gene has two allelic forms, dominant A and recessive a, and there are three possible genotypes- AA (homozygous dominant), Aa (heterozygous), and aa (homozygous recessive).

It is a generally accepted theory that inherited genotype, transmitted epigenetic factors, and non-hereditary environmental variation contribute to the phenotype of an individual.

Non-hereditary DNA mutations are not classically understood as representing the individuals' genotype. Hence, scientists and doctors sometimes talk for example about the (geno)type of a particular cancer, that is the genotype of the disease as distinct from the diseased.

Genotype and Genomic Sequence

One's genotype differs subtly from one's genomic sequence. A sequence is not an absolute measure of base composition of an individual, or a representative of a species or group; a genotype typically implies a measurement of how an individual differs or is specialized within a group of individuals or a species. So typically, one refers to an individual's genotype with regard to a particular gene of interest and, in polyploid individuals, it refers to what combination of alleles the individual carries.

Genotype and Phenotype

Any given gene will usually cause an observable change in an organism, known as the phenotype. The terms genotype and phenotype are distinct for at least two reasons:

1. To distinguish the source of an observer's knowledge (one can know about genotype by observing DNA; one can know about phenotype by observing outward appearance of an organism).
2. Genotype and phenotype are not always directly correlated. Some genes only express a given phenotype in certain environmental conditions. Conversely, some phenotypes could be the result of multiple genotypes. The genotype is commonly mixed up with the Phenotype which describes the end result of both the genetic and the environmental factors giving the observed expression (e.g. blue eyes, hair colour, or various hereditary diseases).

A simple example to illustrate genotype as distinct from phenotype is the flower colour in pea plants (see Gregor Mendel). There are three available genotypes, PP (homozygous dominant), Pp (heterozygous), and pp (homozygous recessive).

All three have different genotypes but the first two have the same phenotype (purple) as distinct from the third (white).

A more technical example to illustrate genotype is the single nucleotide polymorphism or SNP. A SNP occurs when corresponding sequences of DNA from different individuals differ at one DNA base, for example where the sequence AAGCCTA changes to AAGCTTA. This contains two alleles: C and T. SNPs typically have three genotypes, denoted generically AA Aa and aa. In the example above, the three genotypes would be CC, CT and TT. Other types of genetic marker, such as microsatellites, can have more than two alleles, and thus many different genotypes.

Genotype and Mendelian Inheritance

The distinction between genotype and phenotype is commonly experienced when studying family patterns for certain hereditary diseases or conditions, for example, haemophilia. Due to the diploidy of humans (and most animals), there are two alleles for any given gene. These alleles can be the same (homozygous) or different (heterozygous), depending on the individual. With a dominant allele, the offspring is guaranteed to inherit the trait in question irrespective of the second allele. With a recessive allele, the phenotype depends upon the other allele. In the case of haemophilia and similarly recessive diseases a heterozygous individual is a carrier. This person has a normal phenotype but runs a 50-50 risk of passing his or her abnormal gene on to offspring. A homozygous dominant individual has a normal phenotype and no risk of abnormal offspring. A homozygous recessive individual has an abnormal phenotype and is guaranteed to pass the abnormal gene onto offspring.

Genotype and Genetics

With careful experimental design, one can use statistical methods to correlate differences in the genotypes of populations with differences in their observed phenotype. These genetic association studies can be used to determine the genetic risk factors associated with a disease. They may even be able to differentiate between populations who may or may not respond favorably to a particular drug treatment. Such an approach is known as personalized medicine or pharmacogenetics.

Genotype and Mathematics

Inspired by the biological concept and usefulness of genotypes, computer science employs simulated phenotypes in genetic programming and evolutionary algorithms. Such techniques can help evolve mathematical solutions to certain types of otherwise difficult problems.

Determining Genotype

Genotyping is the process of elucidating the genotype of an individual with a biological assay. Also known as a genotypic assay, techniques include PCR, DNA fragment analysis, ASO probes, sequencing, and nucleic acid hybridization to microarrays or beads. Several common genotyping techniques include Restriction Fragment Length Polymorphism (RFLP), Terminal Restriction Fragment Length Polymorphism (t-RFLP) Amplified Fragment Length Polymorphisms (AFLP) and Multiplex Ligation-dependent Probe Amplification (MLPA). DNA fragment analysis can also be used to determine such disease causing genetics aberrations as Microsatellite Instability (MSI), Trisomy or Aneuploidy, and Loss of Heterozygosity (LOH) MSI and LOH in particular have been associated with cancer cell genotypes for colon, breast, and cervical cancer. The most common chromosomal aneuploidy is a trisomy of chromosome 21 which manifests itself as Down Syndrome. Current technological limitations typically allow only a fraction of an individual's genotype to be determined efficiently.

Chapter–6

EPIGENETICS

INTRODUCTION

In biology, the term epigenetics refers to heritable changes in phenotype (appearance) or gene expression caused by mechanisms other than changes in the underlying DNA sequence (hence the name *epi* - "in addition to" - genetics). These changes may remain through cell divisions for the remainder of the cell's life and may also last for multiple generations. However, there is no change in the underlying DNA sequence of the organism instead, non-genetic factors cause the organism's genes to behave (or "express themselves") differently The best example of epigenetic changes in eukaryotic biology is the process of cellular differentiation. During morphogenesis, totipotent stem cells become the various pluripotent cell lines of the embryo which in turn become fully differentiated cells. In other words, a single fertilized egg cell - the zygote - changes into the many cell types including neurons, muscle cells, epithelium, blood vessels et cetera as it continues to divide. It does so by activating some genes while inhibiting others.

Etymology and Definitions

The word epigenetics has had many definitions, and much of the confusion surrounding its usage relates to these definitions having changed over time. Initially it was used in a broader, less specific sense but it has become more narrowly linked to specific molecular phenomena occurring in organisms.

Epigenesis is an older word to describe the differentiation of cells from their initial totipotent state in embryonic development. When Waddington coined the term the physical nature of genes and their role in heredity was not known; he used it as a conceptual model of how genes might interact with their surroundings to produce a phenotype.

Robin Holliday defined epigenetics as "the study of the mechanisms of temporal and spatial control of gene activity during the development of complex organisms." Thus epigenetic can be used to describe any aspect other than DNA sequence that influences the development of an organism.

The modern usage of the word is more narrow, referring to heritable traits (over rounds of cell division and sometimes transgenerationally) that do not involve changes to the underlying DNA sequence. The Greek prefix epi- in epigenetics implies features that are "on top of" or "in addition to" genetics; thus epigenetic traits exist on top of or in addition to the traditional molecular basis for inheritance.

The similarity of the word to "genetics" has generated many parallel usages. The "epigenome" is a parallel to the word "genome," and refers to the overall epigenetic state of a cell. The phrase "genetic code" has also been adapted—the "epigenetic code" has been used to describe the set of epigenetic features that create different phenotypes in different cells. Taken to its extreme, the "epigenetic code" could represent the total state of the cell, with the position of each molecule accounted for; more typically, the term is used in reference to systematic efforts to measure specific, relevant forms of epigenetic information such as the histone code or DNA methylation patterns.

Genetic was also used by the psychologist Erik Erikson in his Psychosocial development theory, however that usage is of primarily historical interest.

MOLECULAR BASIS OF EPIGENETICS

The molecular basis of epigenetics is complex. It involves modifications of the activation of certain genes, but not the basic structure of DNA. Additionally, the chromatin proteins associated with DNA may be activated or silenced. What this means is that every cell in your body has the same instruction manual, but different cell types are using different chapters. For example, your neurons contain the DNA instructions to make your fingernails, but for these and most other cells in the body, the necessary genes are turned off. Epigenetic changes are preserved when cells divide. Most epigenetic changes only occur within the course of one individual organism's lifetime, but some epigenetic changes are inherited from one generation to the next Specific epigenetic processes include paramutation, bookmarking, imprinting, gene silencing, X chromosome inactivation, position effect, reprogramming, transvection, maternal effects, the progress of carcinogenesis, many effects of teratogens, regulation of histone modifications and heterochromatin, and technical limitations affecting parthenogenesis and cloning.

Epigenetic research uses a wide range of molecular biologic techniques to further our understanding of epigenetic phenomena, including chromatin immunoprecipitation (together with its large-scale variants ChIP-on-chip and ChIP-seq), fluorescent in situ hybridization, methylation-sensitive restriction enzymes, DNA adenine methyltransferase identification (DamID) and bisulfite sequencing. Furthermore, the use of bioinformatic methods is playing an increasing role (computational epigenetics).

Mechanisms

Several types of epigenetic inheritance systems may play a role in what has become known as cell memory.

DNA Methylation and Chromatin Remodelling

The DNA associates with histone proteins to form chromatin. Because the phenotype of a cell or individual is affected by which of its genes are transcribed, heritable transcription states can give rise to epigenetic effects. There are several layers of regulation of gene expression. One way that genes are regulated is through the remodeling of chromatin. Chromatin is the complex of DNA and the histone proteins with which it associates. Histone proteins are little spheres that DNA wraps around. If the way that DNA is wrapped around the histones changes, gene expression can change as well. Chromatin remodeling is initiated by one of two things:

1. The first way is post translational modification of the amino acids that make up histone proteins. Histone proteins are made up of long chains of amino acids. If you change the amino acids that are in the chain, you can change the shape of the histone sphere. DNA is not completely unwound during replication. It is possible, then, that the modified histones may be carried into each new copy of the DNA. Once there, these histones may act as templates, initiating the surrounding new histones to be shaped in the new way. By altering the shape of the histones around it, these modified histones would ensure that a differentiated cell would STAY differentiated, and not convert back into being a stem cell.

2. The second way is the addition of methyl groups to the DNA, at CpG sites, to convert cytosine to 5-methylcytosine. Cytosine is the nucleotide that our cells can "read". Our cells cannot "read" methylcytosine. If you think of your DNA as an instruction manual again, changing cytosine to methylcytosine is like changing the font on your word document to "wingdings." Since the cell can no longer "read" the gene, the gene is turned off.

The way that the cells stay differentiated in the case of DNA methylation is more clear to us than it is in the case of histone shape. Basically, certain enzymes (such as Dnmt1) "prefer" the methylated cytosine. If this enzyme comes to a "hemimethylated" portion of DNA (DNA where only one strand contains the methylcytosine, and the other side still contains cytosine) the enzyme will methylate the other half.

Although modifications occur throughout the histone sequence, the unstructured termini of histones (called histone tails) are particularly highly modified. These modifications include acetylation, methylation and ubiquitylation. Acetylation is the most highly studied of these modifications. For example, acetylation of the K14 and K9 lysines of the tail of histone H3 by histone acetyltransferase enzymes (HATs) is generally correlated with transcriptional competence.

One mode of thinking is that this tendency of acetylation to be associated with "active" transcription is biophysical in nature. Because lysine normally has a positive charge on the nitrogen at its end, lysine can bind the negatively charged phosphates of the DNA backbone and prevent them from repelling each other. The acetylation event converts the positively charged amine group on the side chain into a neutral amide linkage. This removes the positive charge causing the DNA to repel itself. When this occurs, complexes like SWI/SNF and other transcriptional factors can bind to the DNA, thus opening it up and exposing it to enzymes like RNA polymerase so transcription of the gene can occur.

In addition, the positively charged tails of histone proteins from one nucleosome may interact with the histone proteins on a neighboring nucleosome, causing them to pack closely. Lysine acetylation may interfere with these interactions, causing the chromatin structure to open up.

Lysine acetylation may also act as a beacon to recruit other activating chromatin modifying enzymes (and basal transcription machinery as well). Indeed, the bromodomain—a protein segment (domain) that specifically binds acetyl-lysine—is found in many enzymes that help activate transcription including the SWI/SNF complex (on the protein polybromo). It may be that acetylation acts in this and the previous way to aid in transcriptional activation.

The idea that modifications act as docking modules for related factors is borne out by histone methylation as well. Methylation of lysine 9 of histone H3 has long been associated with constitutively transcriptionally silent chromatin (constitutive heterochromatin). It has been determined that a chromodomain (a domain that specifically binds methyl-lysine) in the transcriptionally repressive protein HP1 recruits HP1 to K9 methylated regions. One example that seems to refute the biophysical model for acetylation is that tri-methylation of histone H3 at lysine 4 is strongly associated with (and required for full) transcriptional activation. Tri-methylation in this case would introduce a fixed positive charge on the tail.

It should be emphasized that differing histone modifications are likely to function in differing ways; acetylation at one position is likely to function differently than acetylation at another position. Also, multiple modifications may occur at the same time, and these modifications may work together to change the behavior of the nucleosome. The idea that multiple dynamic modifications regulate gene transcription in a systematic and reproducible way is called the histone code.

The DNA methylation frequently occurs in repeated sequences, and may help to suppress 'junk DNA': Because 5-methylcytosine is chemically very similar to thymidine, CpG sites are frequently mutated and become rare in the genome, except at CpG islands where they remain unmethylated. Epigenetic changes of this type thus have the potential to direct increased frequencies of permanent genetic mutation. The DNA methylation patterns are known to be established and modified in response to environmental factors by a complex interplay of at least three independent DNA methyltransferases, DNMT1, DNMT3A and DNMT3B, the loss of any of which is lethal in mice. DNMT1 is the most abundant methyltransferase in somatic cells localizes to replication foci,[has a 10–40-fold preference for hemimethylated DNA and interacts with the proliferating cell nuclear antigen (PCNA) By preferentially modifying hemimethylated DNA, DNMT1 transfers patterns of methylation to a newly synthesized strand after DNA replication, and therefore is often referred to as the 'maintenance' methyltransferase DNMT1 is essential for proper embryonic development, imprinting and X-inactivation.

Chromosomal regions can adopt stable and heritable alternative states resulting in bistable gene expression without changes to the DNA sequence. Epigenetic control is often associated with alternative covalent modifications of histones. The stability and heritability of states of larger chromosomal regions are often thought to involve positive feedback where modified nucleosomes recruit enzymes that similarly modify nearby nucleosomes. A simplified stochastic model for this type of epigenetics is found here.

Because DNA methylation and chromatin remodeling play such a central role in many types of epigenic inheritance, the word "epigenetics" is sometimes used as a synonym for these processes. However, this can be misleading. Chromatin remodeling is not always inherited, and not all epigenetic inheritance involves chromatin remodeling.

It has been suggested that the histone code could be mediated by the effect of small RNAs. The recent discovery and characterization of a vast array of small (21- to 26-nt), non-coding RNAs suggests that there is an RNA component, possibly involved in epigenetic gene regulation. Small interfering RNAs can modulate transcriptional gene expression via epigenetic modulation of targeted promoters.

RNA TRANSCRIPTS AND THEIR ENCODED PROTEINS

Sometimes a gene, after being turned on, transcribes a product that (either directly or indirectly) maintains the activity of that gene. For example, Hnf4 and MyoD enhance the transcription of many liver- and muscle-specific genes, respectively, including their own, through the transcription factor activity of the proteins they encode. Other epigenetic changes are mediated by the production of different splice forms of RNA, or by formation of double-stranded RNA (RNAi). Descendants of the cell in which the gene was turned on will inherit this activity, even if the original stimulus for gene-activation is no longer present. These genes are most often turned on or off by signal transduction, although in some systems where syncytia or gap junctions are important, RNA may spread directly to other cells or nuclei by diffusion.

A large amount of RNA and protein is contributed to the zygote by the mother during oogenesis or via nurse cells, resulting in maternal effect phenotypes. A smaller quantity of sperm RNA is transmitted from the father, but there is recent evidence that this epigenetic information can lead to visible changes in several generations of offspring

Prions

Prions are infectious forms of proteins. Proteins generally fold into discrete units which perform distinct cellular functions, but some proteins are also capable of forming an infectious conformational state known as a prion. Although often viewed in the context of infectious disease, prions are more loosely defined by their ability to catalytically convert other native state versions of the same protein to an infectious conformational state. It is in this latter sense that they can be viewed as epigenetic agents capable of inducing a phenotypic change without a modification of the genome

Fungal prions are considered epigenetic because the infectious phenotype caused by the prion can be inherited without modification of the genome. PSI^+ and URE3, discovered in yeast in 1965 and 1971, are the two best studied of this type of prion Prions can have a phenotypic effect through the sequestration of protein in aggregates, thereby reducing that protein's activity. In PSI^+ cells, the loss of the Sup35 protein (which is involved in termination of translation) causes ribosomes to have a higher rate of read-through of stop codons, an effect which results in suppression of nonsense mutations in other genes. The ability of Sup35 to form prions may be a conserved trait. It could confer an adaptive advantage by giving cells the ability to switch into a PSI^+ state and express dormant genetic features normally terminated by premature stop codon mutations.

Structural Inheritance Systems

In ciliates such as Tetrahymena and Paramecium, genetically identical cells show heritable differences in the patterns of ciliary rows on their cell surface. Experimentally altered patterns can be transmitted to daughter cells. It seems existing structures act as templates for new structures. The mechanisms of such inheritance are unclear, but reasons exist to assume that multicellular organisms also use existing cell structures to assemble new ones

Functions and Consequences Development

Somatic epigenetic inheritance, particularly through DNA methylation and chromatin remodeling, is very important in the development of multicellular eukaryotic organisms. The genome sequence is static (with some notable exceptions), but cells differentiate in many different types, which perform different functions, and respond differently to the environment and intercellular signalling. Thus, as individuals develop, morphogens activate or silence genes in an epigenetically heritable fashion, giving cells a "memory". In mammals, most cells terminally differentiate, with only stem cells retaining the ability to differentiate into several cell types ("totipotency" and "multi-potency"). In mammals, some stem cells continue producing new differentiated cells throughout life, but mammals are not able to respond to loss of some tissues, for example, the inability to regenerate limbs, which some other animals are capable of. Unlike animals, plant cells do not terminally differentiate, remaining totipotent with the ability to give rise to a new individual plant. While plants do utilise many of the same epigenetic mechanisms as animals, such as chromatin remodeling, it has been hypothesised that plant cells do not have "memories", resetting their gene expression patterns at each cell division using positional information from the environment and surrounding cells to determine their fate.

Medicine

Epigenetics has many and varied potential medical applications. Congenital genetic disease is well understood, and it is also clear that epigenetics can play a role, for example, in the case of Angelman syndrome and Prader-Willi syndrome. These are normal genetic diseases caused by gene deletions, but are unusually common because individuals are essentially hemizygous because of genomic imprinting, and therefore a single gene knock out is sufficient to cause the disease, where most cases would require both copies to be knocked out.

Evolution

Although epigenetics in multicellular organisms is generally thought to be a mechanism involved in differentiation, with epigenetic patterns "reset" when organisms reproduce, there have been some observations of transgenerational epigenetic inheritance (e.g., the phenomenon of paramutation observed in maize). Although most of these multi-generational epigenetic traits are gradually lost over several generations, the possibility remains that multi-generational epigenetics could be another aspect to evolution and adaptation. These effects may require enhancements to the standard conceptual framework of the modern evolutionary synthesis.

Epigenetic features may play a role in short-term adaptation of species by allowing for reversible phenotype variability. The modification of epigenetic features associated with a region of DNA allows organisms, on a multigenerational time scale, to switch between phenotypes that express and repress that particular gene. Whereas the DNA sequence of the region is not mutated, this change is reversible. It has also been speculated that organisms may take advantage of differential mutation rates associated with epigenetic features to control the mutation rates of particular genes.

Epigenetic changes have also been observed to occur in response to environmental exposure—for example, mice given some dietary supplements have epigenetic changes affecting expression of the agouti gene, which affects their fur color, weight, and propensity to develop cancer.

EPIGENETIC EFFECTS IN HUMANS

Genomic Imprinting and Related Disorders

Some human disorders are associated with genomic imprinting, a phenomenon in mammals where the father and mother contribute different epigenetic patterns for specific genomic loci in their germ cells The most well-known case of imprinting in human disorders is that of Angelman syndrome and Prader-Willi syndrome—both can be produced by the same genetic mutation, chromosome 15q partial deletion, and the particular syndrome that will develop depends on whether the mutation is inherited from the child's mother or from their father. This is due to the presence of genomic imprinting in the region. Beckwith-Wiedemann syndrome is also associated with genomic imprinting, often caused by abnormalities in maternal genomic imprinting of a region on chromosome II.

Transgenerational Epigenetic Observations

Marcus Pembrey and colleagues also observed that the paternal (but not maternal) grandsons of Swedish boys who were exposed to famine in the 19th century were less likely to die of cardiovascular disease; if food was plentiful then diabetes mortality in the grandchildren increased, suggesting that this was a transgenerational epigenetic inheritance

Cancer and Developmental Abnormalities

A variety of compounds are considered as epigenetic carcinogens—they result in an increased incidence of tumors, but they do not show mutagen activity (toxic compounds or pathogens that cause tumors incident to increased regeneration should also be excluded). Examples include diethylstilbestrol, arsenite, hexachlorobenzene, and nickel compounds.

Many teratogens exert specific effects on the fetus by epigenetic mechanisms While epigenetic effects may preserve the effect of a teratogen such as diethylstilbestrol throughout the life of an affected child, the possibility of birth defects resulting from exposure of fathers or in second and succeeding generations of offspring has generally been rejected on theoretical grounds and for lack of evidence However, a range of male-mediated abnormalities have been demonstrated, and more are likely to exist. FDA label information for Vidaza(tm), a formulation of 5-azacitidine (an unmethylatable analog of cytidine that causes hypomethylation when incorporated into DNA) states that "men should be advised not to father a child" while using the drug, citing evidence in treated male mice of reduced fertility, increased embryo loss, and abnormal embryo development. In rats, endocrine differences were observed in offspring of males exposed to morphine. In mice, second generation effects of diethylstilbesterol have been described occurring by epigenetic mechanisms

EPIGENETICS IN MICRO-ORGANISMS

Bacteria make widespread use of postreplicative DNA methylation for the epigenetic control of DNA-protein interactions. Bacteria make use of DNA adenine methylation (rather than DNA cytosine methylation) as an epigenetic signal. DNA adenine methylation is important in bacteria virulence in organisms such as Escherichia coli, Salmonella, Vibrio, Yersinia, Haemophilus, and Brucella. In Alphaproteobacteria, methylation of adenine regulates the cell cycle and couples gene transcription to DNA replication. In Gammaproteobacteria, adenine methylation provides signals for DNA replication, chromosome segregation, mismatch repair, packaging of bacteriophage, transposase activity and regulation of gene expression.

The yeast prion PSI is generated by a conformational change of a trnslation termination factor, which is then inherited by daughter celular organisms to respond rapidly to environmental stress. Prions can be viewed as epigenetic agents capable of inducing a phenotypic change without modification of the genome.

Gene Expression

Gene expression is the process by which inheritable information from a gene, such as the DNA sequence, is made into a functional gene product, such as protein or RNA.

Several steps in the gene expression process may be modulated, including the transcription step and translation step and the post-translational modification of a protein. Gene regulation gives the cell control over structure and function, and is the basis for cellular differentiation, morphogenesis and the versatility and adaptability of any organism. Gene regulation may also serve as a substrate for evolutionary change, since control of the timing, location, and amount of gene expression can have a profound effect on the functions (actions) of the gene in the organism.Non-protein coding genes (e.g. rRNA genes, tRNA genes) are transcribed, but not translated into protein.

Measurement

The expression of many genes is regulated after transcription (i.e., by microRNAs or ubiquitin ligases), so an increase in mRNA concentration need not always increase expression. In fact, mRNA concentration has been shown to be a poor predictor of resultant protein abundance Nevertheless, mRNA levels can be quantitatively measured by Northern blotting, a process in which a sample of RNA is separated on an agarose gel and hybridized to a radio-labeled RNA probe that is complementary to the target sequence. Northern blotting requires the use of radioactive reagents and can have lower data quality than more modern methods (due to the fact that quantification is done by measuring band strength in an image of a gel), but it is still often used. It does, for example, offer the benefit of allowing the discrimination of alternately spliced transcripts.

A more modern low-throughput approach for measuring mRNA abundance is real-time polymerase chain reaction. (The term RT-PCR is used to refer to both reverse transcription PCR as well as real-time PCR, which is also known as quantitative RT-PCR or quantitative PCR (qPCR). With a carefully constructed standard curve qPCR can produce an absolute measurement such as number of copies of mRNA per nanolitre of homogenized tissue. The lower level of noise in data obtained via qPCR often makes this the method of choice, but the price of the required equipment and reagents can be prohibitive.

In addition to low-throughput methods, transcript levels for many genes at once (expression profiling) can be measured with DNA microarray technology or "tag based" technologies like Serial analysis of gene expression (SAGE) or the more advanced version SuperSAGE, which can provide a relative measure of the cellular concentration of different messenger RNAs. Recent advances in microarray technology allow for the quantification, on a single array, of transcript levels for every known gene in the human genome. The great advantage of tag-based methods is the "open architecture", allowing for the exact measurement of any transcript, known or unknown. Especially SuperSAGE recommends itself therefore also for studying organisms with unknown genomes.

Protein levels themselves can be estimated by a number of means. The most commonly used method is to perform a Western blot against the protein of interest, whereby cellular lysate is separated on a polyacrylamide gel and then probed with an antibody to the protein of interest. The antibody can either be conjugated to a fluorophore or to horseradish peroxidase for imaging or quantification. Another commonly used method for assaying the amount of a particular protein in a cell is to fuse a copy of the protein to a reporter gene such as Green fluorescent protein, which can be directly imaged using a fluorescent microscope. Because it is very difficult to clone a GFP-fused protein into its native location in the genome, however, this method often cannot be used to measure endogenous regulatory mechanisms (GFP-fusions are therefore most often expressed on extra-genomic DNA such as an expression vector). Fusing a target protein to a reporter can also change the protein's behavior, including its cellular localization and expression level.

The pattern of detection of a gene or gene product may be described using terms such as facultative, constitutive, circadian, cyclic, housekeeping, or inducible.

REGULATION OF GENE EXPRESSION

Regulation of gene expression is the cellular control of the amount and timing of appearance of the functional product of a gene. Any step of gene expression may be modulated, from the DNA-RNA transcription step to post-translational modification of a protein. Gene regulation gives the cell control over structure and function, and is the basis for cellular differentiation, morphogenesis and the versatility and adaptability of any organism.

Expression System

An expression system consists, minimally, of a source of DNA and the molecular machinery required to transcribe the DNA into mRNA and translate the mRNA into protein using the nutrients and fuel provided. In the broadest sense, this includes every living cell capable of producing protein from DNA. However, an expression system more specifically refers to a laboratory tool, often artificial in some manner, used for assembling the product of a specific gene or genes. It is defined as the "combination of an expression vector, its cloned DNA, and the host for the vector that provide a context to allow foreign gene function in a host cell, that is, produce proteins at a high level".

In addition to these biological tools, certain naturally observed configurations of DNA (genes, promoters, enhancers, repressors) and the associated machinery itself are referred to as an expression system, as in the simple repressor 'switch' expression system in Lambda phage. It is these natural expression systems that inspire artificial expression systems, (such as the Tet-on and Tet-off expression systems).

Each expression system has distinct advantages and liabilities, and may be named after the host, the DNA source or the delivery mechanism for the genetic material. For example, common expression systems include bacteria (such as E.coli), yeast (such as *S.cerevisiae*), plasmid, artificial chromosomes, phage (such as lambda), cell lines, or virus (such as baculovirus, retrovirus, adenovirus).

Over Expression

In the laboratory, the protein encoded by a gene is sometimes expressed in increased quantity. This can come about by increasing the number of copies of the gene or increasing the binding strength of the promoter region.

Often, the DNA sequence for a protein of interest will be cloned or subcloned into a plasmid containing the lac promoter, which is then transformed into the bacterium Escherichia coli. Addition of IPTG (a lactose analog) causes the bacteria to express the protein of interest. However, this strategy does not always yield functional protein, in which case, other organisms or tissue cultures may be more effective. For example, the yeast Saccharomyces cerevisiae is often preferred to bacteria for proteins that undergo extensive posttranslational modification. Nonetheless, bacterial expression has the advantage of easily producing large amounts of protein, which is required for X-ray crystallography or nuclear magnetic resonance experiments for structure determination.

Gene Networks and Expression

Genes have sometimes been regarded as nodes in a network, with inputs being proteins such as transcription factors, and outputs being the level of gene expression. The node itself performs a function, and the operation of these functions have been interpreted as performing a kind of information processing within cell and determine cellular behaviour.

Chapter–7

TRANSCRIPTION

INTRODUCTION

Transcription is the synthesis of RNA under the direction of DNA. RNA synthesis, or transcription, is the process of transcribing DNA nucleotide sequence information into RNA sequence information. Both nucleic acid sequences use complementary language, and the information is simply transcribed, or copied, from one molecule to the other. DNA sequence is enzymatically copied by RNA polymerase to produce a complementary nucleotide RNA strand, called messenger RNA (mRNA), because it carries a genetic message from the DNA to the protein-synthesizing machinery of the cell. One significant difference between RNA and DNA sequence is the presence of U, or uracil in RNA instead of the T, or thymine of DNA. In the case of protein-encoding DNA, transcription is the first step that usually leads to the expression of the genes, by the production of the mRNA intermediate, which is a faithful transcript of the gene's protein-building instruction. The stretch of DNA that is transcribed into an RNA molecule is called a transcription unit. A DNA transcription unit that is translated into protein contains sequences that direct and regulate protein synthesis in addition to coding the sequence that is translated into protein. The regulatory sequence that is before (upstream (-), towards the 3' DNA end) the coding sequence is called 3' untranslated region (3'UTR), and sequence found following (downstream (+), towards the 5' DNA end) the coding sequence is called 5' untranslated region (5'UTR). Transcription has some proofreading mechanisms, but they are fewer and less effective than the controls for copying DNA; therefore, transcription has a lower copying fidelity than DNA replication.

As in DNA replication, RNA is synthesized in the 5' → 3' direction (from the point of view of the growing RNA transcript). Only one of the two DNA strands is transcribed. This strand is called the template strand, because it provides the template for ordering the sequence of nucleotides in an RNA transcript. The other strand is called the coding strand, because its sequence is the same as the newly created RNA transcript (except for uracil being substituted for thymine). The DNA template strand is read 3' → 5' by RNA polymerase and the new RNA strand is synthesized in the 5' → 3' direction.

A polymerase binds to the 3' end of a gene (promoter) on the DNA template strand and travels toward the 5' end.

Transcription is divided into 5 stages: pre-initiation, initiation, promoter clearance, elongation and termination.

PRE-INITIATION

Unlike DNA replication, transcription does not need a primer to start. RNA polymerase simply binds to the DNA and, along with other cofactors, unwinds the DNA to create an initiaccess to the single-stranded DNA template. However, RNA Polymerase does require a promoter like the ation bubble so that the RNA polymerase has sequence.

Proximal (core) Promoters: TATA promoters are found around -10 and -35 bp to the start site of transcription. Not all genes have TATA box promoters and there exists TATA-less promoters as well. The TATA promoter consensus sequence is TATA(A/T)A(A/T). Some strong promoters have UP sequences involved so that the certain RNA polymerases can bind in greater frequencies.

The following are the steps involved in TATA Promoter Complex formation: 1. General transcription factors bind 2. TFIID, TFIIA, TFIIB, TFIIF (w/RNA Polymerase), TFIIH/E The complex is called the closed pre-initiation complex and is closed. Once the structure is opened by TFIIH initiation starts.

Initiation

In bacteria, transcription begins with the binding of RNA polymerase to the promoter in DNA. The RNA polymerase is a core enzyme consisting of five subunits: 2 α subunits, 1 ß subunit, 1 ß' subunit, and 1 ω subunit. At the start of initiation, the core enzyme is associated with a sigma factor (number 70) that aids in finding the appropriate -35 and -10 basepairs downstream of promoter sequences.

Transcription initiation is far more complex in eukaryotes, the main difference being that eukaryotic polymerases do not directly recognize their core promoter sequences. In eukaryotes, a collection of proteins called transcription factors mediate the binding of RNA polymerase and the initiation of transcription. Only after certain transcription factors are attached to the promoter does the RNA polymerase bind to it. The completed assembly of transcription factors and RNA polymerase bind to the promoter, called transcription initiation complex. Transcription in archaea is similar to transcription in eukaryotes.

Once the complex has been opened, initiation starts and the first phosphodiester bond is formed. This is the end of initiation.

Promoter Clearance

After the first bond is synthesized the RNA polymerase must clear the promoter. During this time there is a tendency to release the RNA transcript and produce truncated transcripts. This is called abortive initiation and is common for both eukaryotes and prokaroytes. Once the transcript reaches approximately 23 nucleotides it no longer slips and elongation can occur. This is an ATP dependent process.

Promoter clearance also coincides with phosphorylation of serine 5 on the carboxy terminal domain which is phosphorylated by TFIIH.

ELONGATION

Simple diagram of transcription elongationOne strand of DNA, the template strand (or non-coding strand), is used as a template for RNA synthesis. As transcription proceeds, RNA polymerase traverses the template strand and uses base pairing complementarity with the DNA template to create an RNA copy. Although RNA polymerase traverses the template strand from 3' → 5', the coding (non-template) strand is usually used as the reference point, so transcription is said to go from 5' → 3'. This produces an RNA molecule from 5' → 3', an exact copy of the coding strand (except that thymines are replaced with uracils, and the nucleotides are composed of a ribose (5-carbon) sugar where DNA has deoxyribose (one less oxygen atom) in its sugar-phosphate backbone).

Unlike DNA replication, mRNA transcription can involve multiple RNA polymerases on a single DNA template and multiple rounds of transcription (amplification of particular mRNA), so many mRNA molecules can be produced from a single copy of a gene. This step also involves a proofreading mechanism that can replace incorrectly incorporated bases.

Prokaryotic elongation starts with the "abortive initiation cycle". During this cycle RNA Polymerase will synthesize mRNA fragments 2-12 nucleotides long. This continues to occur until the s factor rearranges, which results in the transcription elongation complex (which gives a 35 bp moving footprint). The s factor is released before 80 nucleotides of mRNA are synthesized.

In Eukaryotic transcription the polymerase can experience pauses. These pauses may be intrinsic to the RNA polymerase or due to chromatin structure. Often the polymerase pauses to allow appropriate RNA editing factors to bind.

Termination

Simple diagram of transcription terminationBacteria use two different strategies for transcription termination: in Rho-independent transcription termination, RNA transcription stops when the newly synthesized RNA molecule forms a G-C rich hairpin loop, followed by a run of U's, which makes it detach from the DNA template. In the "Rho-dependent" type of termination, a protein factor called "Rho" destabilizes the interaction between the template and the mRNA, thus releasing the newly synthesized mRNA from the elongation complex. Transcription termination in eukaryotes is less well understood. It involves cleavage of the new transcript, followed by template-independent addition of As at its new 3' end, in a process called polyadenylation.

Measuring and Detecting Transcription

Transcription can be measured and detected in a variety of ways:

Nuclear Run-on assay, measures the relative abundance of newly formed transcripts.

RNase protection assay and ChIP-Chip of RNAP, detect active transcription sites.

RT-PCR, measures the absolute abundance of total or nuclear RNA levels, which may however differ from transcription rates.

DNA microarrays measures the relative abundance of the global total or nuclear RNA levels, which may however differ from transcription rates.

In situ hybridization, detects the presence of a transcript.

Transcription Factors

Active transcription units are clustered in the nucleus, in discrete sites called 'transcription factors'. Such sites could be visualized after allowing engaged polymerases to extend their transcripts in tagged precursors (Br-UTP or Br-U), and immuno-labeling the tagged nascent RNA. Transcription factors can also be localized using fluorescence in situ hybridization, or marked by antibodies directed against polymerases. There are ~10,000 factors in the nucleoplasm of a HeLa cell, among which are ~8,000 polymerase II factors and ~2,000 polymerase III factors. Each polymerase II factor contains ~8 polymerases. As most active transcription units are associated with only one polymerase, each factor will be associated with ~8 different transcription units. These units might be associated through promoters and/or enhancers, with loops forming a 'cloud' around the factor.

History

A molecule which allows the genetic material to be realized as a protein was first hypothesized by Jacob and Monod. RNA synthesis by RNA polymerase was established *in vitro* by several laboratories by 1965; however, the RNA synthesized by these enzymes had properties that suggested the existence of an additional factor needed to terminate transcription correctly.

Reverse Transcription

Some viruses (such as HIV, the cause of AIDS), have the ability to transcribe RNA into DNA. HIV has an RNA genome that is duplicated into DNA. The resulting DNA can be merged with the DNA genome of the host cell. The main enzyme responsible for synthesis of DNA from an RNA template is called reverse transcriptase. In the case of HIV, reverse transcriptase is responsible for synthesizing a complementary DNA strand (cDNA) to the viral RNA genome. An associated enzyme, ribonuclease H, digests the RNA strand, and reverse transcriptase synthesises a complementary strand of DNA to form a double helix DNA structure. This cDNA is integrated into the host cell's genome via another enzyme (integrase) causing the host cell to generate viral proteins which reassemble into new viral particles. Subsequently, the host cell undergoes programmed cell death (apoptosis).

Some eukaryotic cells contain an enzyme with reverse transcription activity called telomerase. Telomerase is a reverse transcriptase that lengthens the ends of linear chromosomes. Telomerase carries an RNA template from which it synthesizes DNA repeating sequence, or "junk" DNA. This repeated sequence of "junk" DNA is important because every time a linear chromosome is duplicated, it is shortened in length. With "junk" DNA at the ends of chromosomes, the shortening eliminates some repeated, or junk sequence, rather than the protein-encoding DNA sequence that is further away from the chromosome ends. Telomerase is often activated in cancer cells to enable cancer cells to duplicate their genomes without losing important protein-coding DNA sequence. Activation of telomerase could be part of the process that allows cancer cells to become technically immortal.

Ribonucleic Acid (RNA)

Ribonucleic acid (RNA) is a type of molecule that consists of a long chain of nucleotide units. Each nucleotide consists of a nitrogenous base, a ribose sugar, and a phosphate. RNA is very similar to DNA, but differs in a few important structural details: in the cell, RNA is usually single-stranded, while DNA is usually double-stranded; RNA nucleotides contain ribose while DNA contains deoxyribose (a type of ribose that lacks one oxygen atom); and RNA has the base uracil rather than thymine that is present in DNA.

RNA is transcribed from DNA by enzymes called RNA polymerases and is generally further processed by other enzymes. RNA is central to the synthesis of proteins. Here, a type of RNA called messenger RNA carries information from DNA to structures called ribosomes. These ribosomes are made from proteins and ribosomal RNAs, which come together to form a molecular machine that can read messenger RNAs and translate the information they carry into proteins. There are many RNAs with other roles – in particular regulating which genes are expressed, but also as the genomes of most viruses.

Structure

Each nucleotide in RNA contains a ribose sugar, with carbons numbered 1' through 5'. A base is attached to the 1' position, generally adenine (A), cytosine (C), guanine (G) or uracil (U). Adenine and guanine are purines, cytosine and uracil are pyrimidines. A phosphate group is attached to the 3' position of one ribose and the 5' position of the next. The phosphate groups have a negative charge each at physiological pH, making RNA a charged molecule (polyanion). The bases may form hydrogen bonds between cytosine and guanine, between adenine and uracil and between guanine and uracil. However other interactions are possible, such as a group of adenine bases binding to each other in a bulge or the GNRA tetraloop that has a guanine–adenine base-pair

An important structural feature of RNA that distinguishes it from DNA is the presence of a hydroxyl group at the 2' position of the ribose sugar. The presence of this functional group causes the helix to adopt the A-form geometry rather than the B-form most commonly observed in DNA This results in a very deep and narrow major groove and a shallow and wide minor groove. A second consequence of the presence of the 2'-hydroxyl group is that in conformationally flexible regions of an RNA molecule (that is, not involved in formation of a double helix), it can chemically attack the adjacent phosphodiester bond to cleave the backbone.

The RNA is transcribed with only four bases (adenine, cytosine, guanine and uracil) there are numerous modified bases and sugars in mature RNAs. Pseudouridine (γ), in which the linkage between uracil and ribose is changed from a C–N bond to a C–C bond, and ribothymidine (T), are found in various places (most notably in the T γ C loop of tRNA) Another notable modified base is hypoxanthine, a deaminated adenine base whose nucleoside is called inosine. Inosine plays a key role in the wobble hypothesis of the genetic code. There are nearly 100 other naturally occurring modified nucleosides of which pseudouridine and nucleosides with 2'-O-methylribose are the most common The specific roles of many of these modifications in RNA are not fully understood. However, it is notable that in ribosomal RNA, many of the post-transcriptional modifications occur in highly functional regions, such as the peptidyl transferase center and the subunit interface, implying that they are important for normal function.

Secondary structure of a telomerase RNAThe functional form of single stranded RNA molecules, just like proteins, frequently requires a specific tertiary structure. The scaffold for this structure is provided by secondary structural elements which are hydrogen bonds within the molecule. This leads to several recognizable "domains" of secondary structure like hairpin loops, bulges and internal loops Since RNA is charged, metal ions such as Mg^{2+} are needed to stabilise many secondary structures

Comparison with DNA

RNA and DNA are both nucleic acids, but differ in three main ways. First, unlike DNA which is double-stranded, RNA is a single-stranded molecule in most of its biological roles and has a much shorter chain of nucleotides. Second, while DNA contains deoxyribose, RNA contains ribose, (there is no hydroxyl group attached to the pentose ring in the 2' position in DNA). These hydroxyl groups make RNA less stable than DNA because it is more prone to hydrolysis. Third, the complementary base to adenine is not thymine, as it is in DNA, but rather uracil, which is an unmethylated form of thymine.

Like DNA, most biologically active RNAs including tRNA, rRNA, snRNAs and other, non-coding RNAs are extensively base paired to form double helices. Structural analysis of these RNAs have revealed that they are highly structured. Unlike DNA, their structures do not consist of long double helices but rather collections of short helices packed together into structures akin to proteins. In this fashion, RNAs can achieve chemical catalysis, like enzymes. For instance, determination of the structure of the ribosome—an enzyme that catalyzes peptide bond formation—revealed that its active site is composed entirely of RNA.

Synthesis

Synthesis of RNA is usually catalyzed by an enzyme—RNA polymerase—using DNA as a template, a process known as transcription. Initiation of transcription begins with the binding of the enzyme to a promoter sequence in the DNA (usually found "upstream" of a gene). The DNA double helix is unwound by the helicase activity of the enzyme. The enzyme then progresses along the template strand in the 3' to 5' direction, synthesizing a complementary RNA molecule with elongation occurring in the 5' to 3' direction. The DNA sequence also dictates where termination of RNA synthesis will occur.

RNAs are often modified by enzymes after transcription. For example, a poly(A) tail and a 5' cap are added to eukaryotic pre-mRNA and introns are removed by the spliceosome.

There are also a number of RNA-dependent RNA polymerases that use RNA as their template for synthesis of a new strand of RNA. For instance, a number of RNA viruses (such as poliovirus) use this type of enzyme to replicate their genetic material Also, RNA-dependent RNA polymerase is part of the RNA interference pathway in many organisms.

Types of RNA

Overview

Structure of a hammerhead ribozyme, a ribozyme that cuts RNAMessenger RNA (mRNA) is the RNA that carries information from DNA to the ribosome, the sites of protein synthesis (translation) in the cell. The coding sequence of the mRNA determines the amino acid sequence in the protein that is produced. Many RNAs do not code for protein however. These non-coding RNAs can be encoded by their own genes (RNA genes), but can also derive from mRNA introns. The most prominent examples of non-coding RNAs are transfer RNA (tRNA) and ribosomal RNA (rRNA), both of which are involved in the process of translation There are also non-coding RNAs involved in gene regulation, RNA processing and other roles. Certain RNAs are able to catalyse chemical reactions such as cutting and ligating other RNA molecules and the catalysis of peptide bond formation in the ribosome these are known as ribozymes.

IN TRANSLATION

Messenger RNA (mRNA) carries information about a protein sequence to the ribosomes, the protein synthesis factories in the cell. It is coded so that every three nucleotides (a codon) correspond to one amino acid. In eukaryotic cells, once precursor mRNA (pre-mRNA) has been transcribed from DNA, it is processed to mature mRNA. This removes its introns—non-coding sections of the pre-mRNA. The mRNA is then exported from the nucleus to the cytoplasm, where it is bound to ribosomes and translated into its corresponding protein form with the help of tRNA. In prokaryotic cells, which do not have nucleus and cytoplasm compartments, mRNA can bind to ribosomes while it is being transcribed from DNA. After a certain amount of time the message degrades into its component nucleotides with the assistance of ribonucleases.

Transfer RNA (tRNA) is a small RNA chain of about 80 nucleotides that transfers a specific amino acid to a growing polypeptide chain at the ribosomal site of protein synthesis during translation. It has sites for amino acid attachment and an anticodon region for codon recognition that binds to a specific sequence on the messenger RNA chain through hydrogen bonding.

Ribosomal RNA (rRNA) is the catalytic component of the ribosomes. Eukaryotic ribosomes contain four different rRNA molecules: 18S, 5.8S, 28S and 5S rRNA. Three of the rRNA molecules are synthesized in the nucleolus, and one is synthesized elsewhere. In the cytoplasm, ribosomal RNA and protein combine to form a nucleoprotein called a ribosome. The ribosome binds mRNA and carries out protein synthesis. Several ribosomes may be attached to a single mRNA at any time rRNA is extremely abundant and makes up 80% of the 10 mg/ml RNA found in a typical eukaryotic cytoplasm.

Transfer-messenger RNA (tmRNA) is found in many bacteria and plastids. It tags proteins encoded by mRNAs that lack stop codons for degradation and prevents the ribosome from stalling

Regulatory RNAs

Several types of RNA can downregulate gene expression by being complementary to a part of an mRNA or a gene's

DNA. MicroRNAs (miRNA; 21-22 nt) are found in eukaryotes and act through RNA interference (RNAi), where an effector complex of miRNA and enzymes can break down mRNA which the miRNA is complementary to, block the mRNA from being translated, or accelerate its degradation. While small interfering RNAs (siRNA; 20-25 nt) are often produced by breakdown of viral RNA, there are also endogenous sources of siRNAs siRNAs act through RNA interference in a fashion similar to miRNAs. Some miRNAs and siRNAs can cause genes they target to be methylated, thereby decreasing or increasing transcription of those gene Animals have Piwi-interacting RNAs (piRNA; 29-30 nt) which are active in germline cells and are thought to be a defense against transposons and play a role in gametogenesis. All prokaryotes have CRISPR RNAs, a regulatory system analogous to RNA interference Antisense RNAs are widespread; most downregulate a gene, but a few are activators of transcription One way antisense RNA can act is by binding to an mRNA, forming double-stranded RNA that is enzymatically degraded There are many long noncoding RNAs that regulate genes in eukaryotes, one such RNA is Xist which coats one X chromosome in female mammals and inactivates it.

An mRNA may contain regulatory elements itself, such as riboswitches, in the 5' UTR or 3' UTR; these cis-regulatory elements regulate the activity of that mRNA.

In RNA Processing

Uridine to pseudouridine is a common RNA modification.Many RNAs are involved in modifying other RNAs. Introns are spliced out of pre-mRNA by spliceosomes, which contain several small nuclear RNAs (snRNA), or the introns can be ribozymes that are spliced by themselves. RNA can also be altered by having its nucleotides modified to other nucleotides than A, C, G and U. In eukaryotes, modifications of RNA nucleotides are generally directed by small nucleolar RNAs (snoRNA; 60-300 nt), found in the nucleolus and cajal bodies. snoRNAs associate with enzymes and guide them to a spot on an RNA by basepairing to that RNA. These enzymes then perform the nucleotide modification. rRNAs and tRNAs are extensively modified, but snRNAs and mRNAs can also be the target of base modification.

RNA Genomes

Like DNA, RNA can carry genetic information. RNA viruses have genomes composed of RNA, plus a variety of proteins encoded by that genome. The viral genome is replicated by some of those proteins, while other proteins protect the genome as the virus particle moves to a new host cell. Viroids are another group of pathogens, but they consist only of RNA, do not encode any protein and are replicated by a host plant cell's polymerase

In Reverse Transcription

Reverse transcribing viruses replicate their genomes by reverse transcribing DNA copies from their RNA; these DNA copies are then transcribed to new RNA. Retrotransposons also spread by copying DNA and RNA from one another, and telomerase contains an RNA that is used as template for building the ends of eukaryotic chromosomes

Double-stranded RNA

Double-stranded RNA (dsRNA) is RNA with two complementary strands, similar to the DNA found in all cells. dsRNA forms the genetic material of some viruses (double-stranded RNA viruses). Double-stranded RNA such as viral RNA or siRNA can trigger RNA interference in eukaryotes, as well as interferon response in vertebrates.

Discovery

Nucleic acids were discovered in 1868 by Friedrich Miescher, who called the material 'nuclein' since it was found in the nucleus. It was later discovered that prokaryotic cells, which do not have a nucleus, also contain nucleic acids. The role of RNA in protein synthesis was suspected already in 1939. Severo Oc arl Woese realized RNA can be catalytic and proposed that the earliest forms of life relied on RNA both to carry genetic information and to catalyze biochemical reactions—an RNA world In 1976, Walter Fiers and his team determined the first complete nucleotide sequence of an RNA virus genome, that of bacteriophage MS2 In 1990 it was found in petunia that introduced genes can silence similar genes of the plant's own, now known to be a result of RNA interference At about the same time, 22 nt long RNAs, now called microRNAs, were found to have a role in the development of C. elegans. The discovery of gene regulatory RNAs has led to attempts to develop drugs made of RNA, such as siRNA, to silence genes.

Reverse Transcriptase

DNA polymerase, is a DNA polymerase enzyme that transcribes single-stranded RNA into double-stranded DNA. It also helps in the formation of a double helix DNA once the RNA has been reverse transcribed into a single strand cDNA. Normal transcription involves the synthesis of RNA from DNA; hence, reverse transcription is the reverse of this.

Transcriptase was discovered by Howard Temin at the University of Wisconsin-Madison, and independently by David Baltimore in 1970 at MIT. The two shared the 1975 Nobel Prize in Physiology or Medicine with Renato Dulbecco for their discovery.

Well studied reverse transcriptases include:

HIV-1 reverse transcriptase from human immunodeficiency virus type 1 (PDB 1HMV)

M-MLV reverse transcriptase from the Moloney murine leukemia virus

AMV reverse transcriptase from the avian myeloblastosis virus

Telomerase reverse transcriptase that maintains the telomeres of eukaryotic chromosomes

Function

Viruses

The enzyme is encoded and used by reverse-transcribing viruses, which use the enzyme during the process of replication.

Reverse-transcribing RNA viruses, such as retroviruses, use the enzyme to reverse-transcribe their RNA genomes into DNA, which is then integrated into the host genome and replicated along with it. Reverse-transcribing DNA viruses, such as the hepadnaviruses, can allow RNA to serve as a template in assembling, and making DNA strands. HIV infects humans with the use of this enzyme. Without reverse transcriptase, the viral genome would not be able to incorporate into the host cell, resulting in the failure of the ability to replicate. Unlike bacteria, retroviruses use preexisting host-encoded transfer RNAs as primers.

Process of Reverse Transcription

Retroviral RNA is arranged in 5' terminus to 3' terminus. The site where the primer is annealed to viral RNA is called the primer-binding site (PBS). The RNA 5'end to the PBS site is called U5, and the RNA 3' end to the PBS is called the leader. The tRNA primer is unwound between 14 and 22 nucleotides and forms a base-paired duplex with the viral RNA at PBS. PBS locates near the 5' terminus of viral RNA is unusual because reverse transcriptase synthesize DNA from 3' end of the primer in the 5' to 3' direction. Therefore, the primer and reverse transcriptase must be relocated to 3' end of viral RNA. In order to accomplish this reposition, multiple steps and various enzymes included DNA polymerase, ribonuclease H(RNase H) and polynucleotide unwinding are needed.

The reverse transcription of retroviral RNA is a discontinuous process. The production of DNA begins with short and discrete segments which will be elongated during a process called strand transfer. Strand transfer involves translocation of short DNA product from initial synthesis site to acceptor template regions at the other end of the genome. Thus, the final product of reverse transcriptase is a linear double-stranded DNA.

Eukaryotes

Self-replicating stretches of eukaryotic genomes known as retrotransposons utilize reverse transcriptase to move from one position in the genome to another via a RNA intermediate. They are found abundantly in the genomes of plants and animals. Telomerase is another reverse transcriptase found in many eukaryotes, including humans, which carries its own RNA template; this RNA is used as a template for DNA replication.

Prokaryotes

Reverse transcriptases are also found in bacterial Retron msr RNAs, distinct sequences which code for reverse transcriptase, and are used in the synthesis of msDNA. In order to initiate synthesis of DNA, a primer is needed. In bacteria, the primer is synthesized during replication

Structure

Reverse transcriptase enzymes include an RNA-dependent DNA polymerase and a DNA-dependent DNA polymerase, which work together to perform transcription. In addition to the transcription function, retroviral reverse transcriptases have a domain belonging to the RNase H family which is vital to their replication.

Replication Fidelity

There are three different replication systems during the life cycle of a retrovirus. First of all, the reverse transcriptase synthesize viral DNA from viral RNA, and then from newly made complementary DNA strand. The second replication process occurs when host cellular DNA polymerase replicates the integrated viral DNA. Lastly, RNA polymerase II transcribes the proviral DNA into RNA which will be packed into virions. Therefore, mutation can occur during one or all of these replication steps.

Reverse transcriptase has a high error rate when transcribing RNA into DNA since, unlike DNA polymerases, it has no proofreading ability. This high error rate allows mutations to accumulate at an accelerated rate relative to proofread forms of replication. The commercially available reverse transcriptases produced by Promega are quoted by their manuals as having error rates in the range of 1 in 17,000 bases for AMV and 1 in 30,000 bases for M-MLV.

Applications

The molecular structure of zidovudine (AZT), a drug used to inhibit HIV reverse transcriptase

Antiviral Drugs

As HIV uses reverse transcriptase to copy its genetic material and generate new viruses (part of a retrovirus proliferation circle), specific drugs have been designed to disrupt the process and thereby suppress its growth. Collectively, these drugs are known as reverse transcriptase inhibitors and include the nucleoside and nucleotide analogues zidovudine (trade name Retrovir), lamivudine (Epivir) and tenofovir (Viread), as well as non-nucleoside inhibitors, such as nevirapine (Viramune).

Molecular Biology

Reverse transcriptase is commonly used in research to apply the polymerase chain reaction technique to RNA in a technique called reverse transcription polymerase chain reaction (RT-PCR). The classical PCR technique can be applied only to DNA strands, but, with the help of reverse transcriptase, RNA can be transcribed into DNA, thus making PCR analysis of RNA molecules possible. Reverse transcriptase is used also to create cDNA libraries from mRNA. The commercial availability of reverse transcriptase greatly improved knowledge in the area of molecular biology, as, along with other enzymes, it allowed scientists to clone, sequence, and characterise DNA.

DNA POLYMERASE

A DNA polymerase is an enzyme that catalyzes the polymerization of deoxyribonucleotides into a DNA strand. DNA polymerases are best-known for their role in DNA replication, in which the polymerase "reads" an intact DNA strand as a template and uses it to synthesize the new strand. The newly-polymerized molecule is complementary to the template strand and identical to the template's original partner strand. DNA polymerases use a magnesium ion for catalytic activity.

Function

DNA polymerase with proofreading ability.DNA polymerase can add free nucleotides to only the 3' end of the newly-forming strand. This results in elongation of the new strand in a 5'-3' direction. No known DNA polymerase is able to begin a new chain (de novo). DNA polymerase can add a nucleotide onto only a preexisting 3'-OH group, and, therefore, needs a primer at which it can add the first nucleotide. Primers consist of RNA and DNA bases with the first two bases always being RNA, and are synthesized by another enzyme called primase. An enzyme known as a helicase is required to unwind DNA from a double-strand structure to a single-strand structure to facilitate replication of each strand consistent with the semiconservative model of DNA replication.

Error correction is a property of some, but not all, DNA polymerases. This process corrects mistakes in newly-synthesized DNA. When an incorrect base pair is recognized, DNA polymerase reverses its direction by one base pair of DNA. The 3'->5' exonuclease activity of the enzyme allows the incorrect base pair to be excised (this activity is known as proofreading). Following base excision, the polymerase can re-insert the correct base and replication can continue.

Variation Across Species

DNA polymerases have highly-conserved structure, which means that their overall catalytic subunits vary, on a whole, very little from species to species. Conserved structures usually indicate important, irreplicable functions of the cell, the maintenance of which provides evolutionary advantages.

Some viruses also encode special DNA polymerases that may selectively replicate viral DNA through a variety of mechanisms. Retroviruses encode an unusual DNA polymerase called reverse transcriptase, which is an RNA-dependent DNA polymerase (RdDp). It polymerizes DNA from a template of RNA.

DNA Polymerase Families

Based on sequence homology, DNA polymerases can be further subdivided into seven different families: A, B, C, D, X, Y, and RT.

Family A

Polymerases contain both replicative and repair polymerases. Replicative members from this family include the extensively-studied T7 DNA polymerase, as well as the eukaryotic mitochondrial DNA Polymerase γ. Among the repair polymerases are *E. coli* DNA pol I, *Thermus aquaticus pol I*, and *Bacillus stearothermophilus* pol I. These repair polymerases are involved in excision repair and processing of Okazaki fragments generated during lagging strand synthesis.

Family B

Polymerases mostly contain replicative polymerases and include the major eukaryotic DNA polymerases α, δ, ε, (see Greek letters) and also DNA polymerase. Family B also includes DNA polymerases encoded by some bacteria and bacteriophages, of which the best-characterized are from T4, Phi29, and RB69 bacteriophages. These enzymes are involved in both leading and lagging strand synthesis. A hallmark of the B family of polymerases is remarkable accuracy during replication; and many have strong 3'-5' exonuclease activity (except DNA polymerase a and *g* which have no proofreading activity).

Family C

Polymerases are the primary bacterial chromosomal replicative enzymes. DNA Polymerase III alpha subunit from *E. coli* is the catalytic subunit and possesses no known nuclease activity. A separate subunit, the epsilon subunit, possesses the 3'-5' exonuclease activity used for editing during chromosomal replication.

Family D

Polymerases are still not very well characterized. All known examples are found in the Euryarchaeota subdomain of Archaea and are thought to be replicative polymerases.

Family X

Family X contains the well-known eukaryotic polymerase pol ß, as well as other eukaryotic polymerases such as pol σ, pol →, pol μ, and terminal deoxynucleotidyl transferase (TdT). Pol ß is required for short-patch base excision repair, a DNA repair pathway that is essential for repairing abasic sites. Pol ? and Pol μ are involved in non-homologous end-joining, a mechanism for rejoining DNA double-strand breaks. TdT is expressed only in lymphoid tissue, and adds "n nucleotides" to double-strand breaks formed during V(D)J recombination to promote immunological diversity. The yeast *Saccharomyces cerevisiae* has only one Pol X polymerase, Pol4, which is involved in non-homologous end-joining.

Family Y

Polymerases differ from others in having a low fidelity on undamaged templates and in their ability to replicate through damaged DNA. Members of this family are hence called translesion synthesis (TLS) polymerases. Depending on the lesion, TLS polymerases can bypass the damage in an error-free or error-prone fashion, the latter resulting in elevated mutagenesis. Xeroderma pigmentosum variant (XPV) patients for instance have mutations in the gene encoding Pol η (eta), which is error-free for UV-lesions. In XPV patients, alternative error-prone polymerases, e.g., Pol ξ (zeta) (polymerase ξ is a B Family polymerase), are thought to be involved in mistakes that result in the cancer predisposition of these patients. Other members in humans are Pol ι (iota), Pol κ (kappa), and Rev1 (terminal deoxycytidyl transferase). In *E.coli*, two TLS polymerases, Pol IV (DINB) and PolV (UmuD'2C), are known.

Family RT

The reverse transcriptase family contains examples from both retroviruses and eukaryotic polymerases. The eukaryotic polymerases are usually restricted to telomerases. These polymerases use an RNA template to synthesize the DNA strand.

Prokaryotic DNA Polymerases

Bacteria have 5 known DNA polymerases:

- *Pol I:* implicated in DNA repair; has both 5'→3'(Polymerase) activity and 5'→3' exonuclease activity (in removing RNA primers).
- *Pol II:* involved in replication of damaged DNA; has 3'→5' exonuclease activity.
- *Pol III:* the main polymerase in bacteria (elongates in DNA replication); has 3'→5' exonuclease proofreading ability.
- *Pol IV:* a Y-family DNA polymerase.
- *Pol V:* a Y-family DNA polymerase; participates in bypassing DNA damage.

Eukaryotic DNA Polymerases

Eukaryotes have at least 15 DNA Polymerases:

- Pol α (synonyms are RNA primase, DNA polymerase): acts as a primase (synthesizing an RNA primer), and then as a DNA Pol elongating that primer with DNA nucleotides. After around 20 nucleotides elongation is taken over by Pol ε (on the lagging strand) and δ (on the leading strand).
- Pol ß: Implicated in repairing DNA, in base excision repair and gap-filling synthesis.
- Pol γ: Replicates and repairs mitochondrial DNA and has proofreading 3'→5' exonuclease activity.
- Pol δ: Highly processive and has proofreading 3'→5' exonuclease activity. Thought to be the main polymerase involved in leading strand synthesis, though there is still debate about its role.
- Pol ε: Also highly processive and has proofreading 3'→5' exonuclease activity. Highly related to pol δ, and thought

to be the main polymerase involved in lagging strand synthesis, though there is again still debate about its role.

η, ι, κ, and Rev1 are Y-family DNA polymerases and Pol ζ is a B-family DNA polymerase. These polymerases are involved in the bypass of DNA damage.

- There are also other eukaryotic polymerases known, which are not as well characterized: θ, λ, φ, σ, and μ. There are also others, but the nomenclature has become quite jumbled.

None of the eukaryotic polymerases can remove primers (5'→3' exonuclease activity); that function is carried out by other enzymes. Only the polymerases that deal with the elongation (γ, δ and ε) have proofreading ability (3'→5' exonuclease). Nuclease is a great advocate for DNA replication.

RNA POLYMERASE

RNA polymerase (RNAP or RNApol) is an enzyme that produces RNA. In cells, RNAP is needed for constructing RNA chains from DNA genes as templates, a process called transcription. RNA polymerase enzymes are essential to life and are found in all organisms and many viruses. In chemical terms, RNAP is a nucleotidyl transferase that polymerizes ribonucleotides at the 3' end of an RNA transcript.

History

RNAP was discovered independently by Sam Weiss and Jerard Hurwitz in 1960 By this time the 1959 Nobel Prize in Medicine had been awarded to Severo Ochoa and Arthur Kornberg for the discovery of what was believed to be RNAP but instead turned out to be polynucleotide phosphorylase.

The 2006 Nobel Prize in Chemistry was awarded to Roger Kornberg for creating detailed molecular images of RNA polymerase during various stages of the transcription process.

Control of Transcription

Control of the process of gene transcription affects patterns of gene expression and thereby allows a cell to adapt to a changing environment, perform specialized roles within an organism, and maintain basic metabolic processes necessary for survival. Therefore, it is hardly surprising that the activity of RNAP is both complex and highly regulated. In *Escherichia coli* bacteria, more than 100 transcription factors have been identified which modify the activity of RNAP.

RNAP can initiate transcription at specific DNA sequences known as promoters. It then produces an RNA chain which is complementary to the template DNA strand. The process of adding nucleotides to the RNA strand is known as elongation; In eukaryotes, RNAP can build chains as long as 2.4 million nucleosides (the full length of the dystrophin gene). RNAP will preferentially release its RNA transcript at specific DNA sequences encoded at the end of genes known as terminators.

Products of RNAP include:

- Messenger RNA (mRNA)—template for the synthesis of proteins by ribosomes.
- Non-coding RNA or "RNA genes"—a broad class of genes that encode RNA that is not translated into protein. The most prominent examples of RNA genes are transfer RNA (tRNA) and ribosomal RNA (rRNA), both of which are involved in the process of translation. However, since the late 1990s, many new RNA genes have been found, and thus RNA genes may play a much more significant role than previously thought.

Transfer RNA (tRNA)—transfers specific amino acids to growing polypeptide chains at the ribosomal site of protein synthesis during translation.

Ribosomal RNA (rRNA)—a component of ribosomes.

Micro RNA—regulates gene activity.

Catalytic RNA (Ribozyme)—enzymatically active RNA molecules.

RNAP accomplishes *de novo* synthesis. It is able to do this because specific interactions with the initiating nucleotide hold RNAP rigidly in place, facilitating chemical attack on the incoming nucleotide. Such specific interactions explain why RNAP prefers to start transcripts with ATP (followed by GTP, UTP, and then CTP). In contrast to DNA polymerase, RNAP includes helicase activity, therefore no separate enzyme is needed to unwind DNA.

RNA Polymerase Action

Binding and Initiation

RNA Polymerase binding in prokaryotes involves the a subunit recognizing the upstream element (-40 to -70 base pairs) in DNA, as well as the s factor recognizing the -10 to -35 region. There are numerous s factors that regulate gene expression. For example, σ^{70} is expressed under normal conditions and allows RNAP binding to house-keeping genes, while σ^{32} elicits RNAP binding to heat-shock genes.

After binding to the DNA, the RNA polymerase switches from a closed complex to an open complex. This change involves the separation of the DNA strands to form an unwound section of DNA of approximately 13bp. Ribonucleotides are base-paired to the template DNA strand, according to Watson-Crick base-pairing interactions. Supercoiling plays an important part in polymerase activity because of the unwinding and rewinding of DNA. Because regions of DNA in front of RNAP are unwound, there is compensatory positive supercoils. Regions behind RNAP are rewound and negative supercoils are present.

Elongation

Transcription elongation involves the further addition of ribonucleotides and the change of the open complex to the transcriptional complex. RNAP cannot start forming full length transcripts because of its strong binding to promoter. Transcription at this stage primarily results in short RNA fragments of around 9 bp in a process known as abortive transcription. Once the RNAP starts forming longer transcripts it clears the promoter. At this point, the -10 to -35 promoter region is disrupted, and the s factor falls off RNAP. This allows the rest of the RNAP complex to move forward, as the σ factor held the RNAP complex in place.

The 17 bp transcriptional complex has an 8 bp DNA-RNA hybrid, that is, 8 base-pairs involve the RNA transcript bound to the DNA template strand. As transcription progresses, ribonucleotides are added to the 3' end of the RNA transcript and the RNAP complex moves along the DNA. Although RNAP does not seem to have the 3'exonuclease activity that characterizes the proof-reading activity found in DNA polymerase, there is evidence of that RNAP will halt at mismatched base-pairs and correct it.

The addition of ribonucleotides to the RNA transcript has a very similar mechanism to DNA polymerization - it is believed that these polymerases are evolutionarily related. Aspartyl (asp) residues in the RNAP will hold onto Mg^{2+} ions, which will in turn coordinate the phosphates of the ribonucleotides. The first Mg^{2+} will hold onto the a-phosphate of the NTP to be added. This allows the nucleophilic attack of the 3'OH from the RNA transcript, adding an additional NTP to the chain. The second Mg^{2+} will hold onto the pyrophosphate of the NTP. The overall reaction equation is:

$$(NMP)n + NTP \rightarrow (NMP)_{n+1} + PPi$$

Termination

Termination of RNA transcription can be rho-independent or rho-dependent:

Rho-independent transcription termination is the termination of transcription without the aid of the rho protein. Transcription of a palindromic region of DNA causes the formation of a hairpin structure from the RNA transcription looping and binding upon itself. This hairpin structure is often rich in G-C base-pairs, making it more stable than the DNA-RNA hybrid itself. As a result, the 8bp DNA-RNA hybrid in the transcription complex shifts to a 4bp hybrid. Coincidentally, these last 4 base-pairs are weak A-U base-pairs, and the entire RNA transcript will fall off.

RNA Polymerase in Bacteria

- In bacteria, the same enzyme catalyzes the synthesis of mRNA and ncRNA.
- RNAP is a relatively large molecule. The core enzyme has 5 subunits (~400 kDa):
- α_2: the two a subunits assemble the enzyme and recognize regulatory factors. Each subunit has two domains: αCTD (C-Terminal domain) binds the UP element of the extended promoter, and αNTD (N-terminal domain) binds the rest of the polymerase. This subunit is not used on promoters without an UP element.
- ß: this has the polymerase activity (catalyzes the synthesis of RNA) which includes chain initiation and elongation.
- ß': binds to DNA (nonspecifically).
- ω: restores denatured RNA polymerase to its functional form *in vitro*. It has been observed to offer a protective/ chaperone function to the ß' subunit in *Mycobacterium* smegmatis. Now known to promote assembly.

In order to bind promoter-specific regions, the core enzyme requires another subunit, sigma (σ). The sigma factor greatly reduces the affinity of RNAP for nonspecific DNA while increasing specificity for certain promoter regions, depending on the sigma factor. That way, transcription is initiated at the right region. The complete holoenzyme therefore has 6 subunits: $\alpha_2\beta\beta'\sigma\omega$ (~480 kDa). The structure of RNAP exhibits a groove with a length of 55 Å (5.5 nm) and a diameter of 25 Å (2.5 nm). This groove fits well the 20 Å (2 nm) double strand of DNA. The 55 Å (5.5 nm) length can accept 16 nucleotides.

When not in use RNA polymerase binds to low affinity sites to allow rapid exchange for an active promoter site when one opens. RNA polymerase holoenzyme, therefore, does not freely float around in the cell when not in use.

Transcriptional Cofactors

There are a number of proteins which can bind to RNAP and modify its behaviour. For instance, GreA and GreB from *E. coli* and in most other prokaryotes can enhance the ability of RNAP to cleave the RNA template near the growing end of the chain. This cleavage can rescue a stalled polymerase molecule, and is likely involved in proofreading the occasional mistakes made by RNAP. A separate cofactor, Mfd, is involved in transcription-coupled repair, the process in which RNAP recognizes damaged bases in the DNA template and recruits enzymes to restore the DNA. Other cofactors are known to play regulatory roles, i.e. they help RNAP choose whether or not to express certain genes.

RNA Polymerase in Eukaryotes

Eukaryotes have several types of RNAP, characterized by the type of RNA they synthesize:

- RNA polymerase I synthesizes a pre-rRNA 45S, which matures into 28S, 18S and 5.8S rRNAs which will form the major RNA sections of the ribosome.
- RNA polymerase II synthesizes precursors of mRNAs and most snRNA and microRNAs This is the most studied type, and due to the high level of control required over transcription a range of transcription factors are required for its binding to promoters.
- RNA polymerase III synthesizes tRNAs, rRNA 5S an RNA polymerase IV synthesizes siRNA in plants.

There are other RNA polymerase types in mitochondria and chloroplasts. And there are RNA-dependent RNA polymerases involved in RNA interference.

RNA Polymerase in Archaea

Archaea have a single RNAP that is closely related to the three main eukaryotic polymerases. Thus, it has been speculated that the archaeal polymerase resembles the ancestor of the specialized eukaryotic polymerases

RNA Polymerase in Viruses

Many viruses also encode for RNAP. Perhaps the most widely studied viral RNAP is found in bacteriophage T7. This single-subunit RNAP is related to that found in mitochondria and chloroplasts, and shares considerable homology to DNA polymerase.It is believed that most viral polymerases therefore evolved from DNA polymerase and are not directly related to the multi-subunit polymerases described above.

The viral polymerases are diverse, and include some forms which can use RNA as a template instead of DNA. This occurs in negative strand RNA viruses and dsRNA viruses, both of which exist for a portion of their life cycle as double-stranded RNA. However, some positive strand RNA viruses, such as polio, also contain these RNA dependent RNA polymerases.

RNA Polymerase Purification

RNA polymerase can be isolated in the following ways:

- By a phosphocellulose column.
- By glycerol gradient centrifugation.
- By a DNA column.
- By a Ion exchange column.

And also combinations of the above techniques.

Chapter–8

Molecular Phylogenetics

INTRODUCTION

Molecular phylogenetics, also known as molecular systematics, is the use of the structure of molecules to gain information on an organism's evolutionary relationships. The result of a molecular phylogenetic analysis is expressed in a phylogenetic tree.

Techniques and Applications

Every living organism contains DNA, RNA, and proteins. Closely related organisms generally have a high degree of agreement in the molecular structure of these substances, while the molecules of organisms distantly related usually show a pattern of dissimilarity. Molecular phylogeny uses such data to build a "relationship tree" that shows the probable evolution of various organisms. Not until recent decades, however, has it been possible to isolate and identify these molecular structures.

The most common approach is the comparison of sequences for genes using sequence alignment techniques to identify similarity. Another application of molecular phylogeny is in DNA barcoding, where the species of an individual organism is identified using small sections of mitochondrial DNA. Another application of the techniques that make this possible can be seen in the very limited field of human genetics, such as the ever more popular use of genetic testing to determine a child's paternity, as well as the emergence of a new branch of criminal forensics focused on evidence known as genetic fingerprinting.

The effect on traditional biological classification schemes in the biological sciences has been dramatic as well. Work that was once immensely labour- and materials-intensive can now be done quickly and easily, leading to yet another source of information becoming available for systematic and taxonomic appraisal. This particular kind of data has become so popular that taxonomical schemes based solely on molecular data may be encountered.

Theoretical Background

Early attempts at molecular systematics were also termed as chemotaxonomy and made use of proteins, enzymes, carbohydrates and other molecules which were separated and characterized using techniques such as chromatography. These have been largely replaced in recent times by DNA sequencing which produces the exact sequences of nucleotides or bases in either DNA or RNA segments extracted using different techniques. These are generally considered superior for evolutionary studies since the actions of evolution are ultimately reflected in the genetic sequences. At present it is still a long and expensive process to sequence the entire DNA of an organism (its genome), and this has been done for only a few species. However it is quite feasible to determine the sequence of a defined area of a particular chromosome. Typical molecular systematic analyses require the sequencing of around 1000 base pairs. At any location within such a sequence, the bases found in a given position may vary between organisms. The particular sequence found in a given organism is referred to as its haplotype. In principle, since there are four base types, with 1,000 base pairs, we could have 41,000 distinct haplotypes. However, for organisms within a particular species or in a group of related species, it has been found empirically that only a minority of sites show any variation at all and most of the variations that are found are correlated, so that the number of distinct haplotypes that are found is relatively small.

In a molecular systematic analysis, the haplotypes are determined for a defined area of genetic material; ideally a substantial sample of individuals of the target species or other taxon are used however many current studies are based on single individuals. Haplotypes of individuals of closely related, but supposedly different, taxa are also determined. Finally, haplotypes from a smaller number of individuals from a definitely different taxon are determined: these are referred to as an out group. The base sequences for the haplotypes are then compared. In the simplest case, the difference between two haplotypes is assessed by counting the number of locations where they have different bases: this is referred to as the number of substitutions (other kinds of differences between haplotypes can also occur, for example the insertion of a section of nucleic acid in one haplotype that is not present in another). Usually the difference between organisms is re-expressed as a percentage divergence, by dividing the number of substitutions by the number of base pairs analysed: the hope is that this measure will be independent of the location and length of the section of DNA that is sequenced.

An older and superseded approach was to determine the divergences between the genotypes of individuals by DNA-DNA hybridisation. The advantage claimed for using hybridisation rather than gene sequencing was that it was based on the entire genotype, rather than on particular sections of DNA. Modern sequence comparison techniques overcome this objection by the use of multiple sequences.

Once the divergences between all pairs of samples have been determined, the resulting triangular matrix of differences is submitted to some form of statistical cluster analysis, and the resulting dendrogram is examined in order to see whether the samples cluster in the way that would be expected from current ideas about the taxonomy of the group, or not. Any group of haplotypes that are all more similar to one another than any of them is to any other haplotype may be said to constitute a clade. Statistical techniques such as bootstrapping and jackknifing help in providing reliability estimates for the positions of haplotypes within the evolutionary trees.

CHARACTERISTICS AND ASSUMPTIONS OF MOLECULAR SYSTEMATICS

This example illustrates several characteristics of molecular systematics and its underlying assumptions:

1. Molecular systematics is an essentially cladistic approach: it assumes that classification must correspond to phylogenetic descent, and that all valid taxa must be monophyletic.
2. Molecular systematics often uses the molecular clock assumption that quantitative similarity of genotype is a sufficient measure of the recency of genetic divergence. Particularly in relation to speciation, this assumption could be wrong if either:
   (a) some relatively small genotypic modification acted to prevent interbreeding between two groups of organisms, or;
   (b) in different subgroups of the organisms being considered, genetic modification proceeded at different rates;
   (c) In animals, it is often convenient to use mitochondrial DNA for molecular systematic analysis. However, because in mammals mitochondria are inherited only from the mother, this is not fully satisfactory, because inheritance in the paternal line might not be detected: in the example above, Vilà et al cite more limited studies with chromosomal DNA that support their conclusions.

These characteristics and assumptions are not wholly uncontroversial among biological systematists. As a cladistic method, molecular systematics is open to the same criticisms as cladistics in general. It can also be argued that it is a mistake to replace a classification based on visible and ecologically relevant characteristics by one based on genetic details that may not even be expressed in the phenotype. However the molecular approach to systematics, and its underlying assumptions, are gaining increasing acceptance. As gene sequencing becomes easier and cheaper, molecular systematics is being applied to more and more groups, and in some cases is leading to radical revisions of accepted taxonomies.

History of Molecular Phylogenetics

Molecular systematics was pioneered by Charles G. Sibley (birds), Herbert C. Dessauer (herpetology), and Morris Goodman (primates), followed by Allan C. Wilson, Robert K. Selander, and John C. Avise (who studied various groups). Work with protein electrophoresis began around 1956. Although the results were not quantitative and did not initially improve on morphological classification, they provided tantalizing hints that long-held notions of the classifications of birds, for example, needed substantial revision. In the period of 1974-1986, DNA-DNA hybridization was the dominant technique.

MOLECULAR EVOLUTION

Molecular evolution is the process of evolution at the scale of DNA, RNA, and proteins. Molecular evolution emerged as a scientific field in the 1960s as researchers from molecular biology, evolutionary biology and population genetics sought to understand recent discoveries on the structure and function of nucleic acids and protein. Some of the key topics that spurred development of the field have been the evolution of enzyme function, the use of nucleic acid divergence as a "molecular clock" to study species divergence, and the origin of non-functional or junk DNA. Recent advances in genomics, including whole-genome sequencing, high-throughput protein characterization, and bioinformatics have led to a dramatic increase in studies on the topic. In the 2000s, some of the active topics have been the role of gene duplication in the emergence of novel gene function, the extent of adaptive molecular evolution versus neutral drift, and the identification of molecular changes responsible for various human characteristics especially those pertaining to infection, disease, and cognition.

PRINCIPLES OF MOLECULAR EVOLUTION

Mutations are permanent, transmissible changes to the genetic material (usually DNA or RNA) of a cell. Mutations can be caused by copying errors in the genetic material during cell division and by exposure to radiation, chemicals, or viruses, or can occur deliberately under cellular control during the processes such as meiosis or hypermutation. Mutations are considered the driving force of evolution, where less favorable (or deleterious) mutations are removed from the gene pool by natural selection, while more favourable (or beneficial) ones tend to accumulate. Neutral mutations do not affect the organism's chances of survival in its natural environment and can accumulate over time, which might result in what is known as punctuated equilibrium; the modern interpretation of classic evolutionary theory.

Causes of Change in Allele Frequency

There are some known processes that affect the survival of a characteristic; or, more specifically, the frequency of an allele (variant of a gene):

Genetic drift describes changes in gene frequency that cannot be ascribed to selective pressures, but are due instead to events that are unrelated to inherited traits. This is especially important in small mating populations, which simply cannot have enough offspring to maintain the same gene distribution as the parental generation.

Gene flow or Migration: or gene admixture is the only one of the agents that makes populations closer genetically while building larger gene pools.

Selection, in particular natural selection produced by differential mortality and fertility. Differential mortality is the survival rate of individuals before their reproductive age. If they survive, they are then selected further by differential fertility – that is, their total genetic contribution to the next generation. In this way, the alleles that these surviving individuals contribute to the gene pool will increase the frequency of those alleles. Sexual selection, the attraction between mates that results from two genes, one for a feature and the other determining a preference for that feature, is also very important.

MOLECULAR STUDY OF PHYLOGENY

Molecular systematics is a product of the traditional field of systematics and molecular genetics. It is the process of using data on the molecular constitution of biological organisms' DNA, RNA, or both, in order to resolve questions in systematics, i.e. about their correct scientific classification or taxonomy from the point of view of evolutionary biology.

Molecular systematics has been made possible by the availability of techniques for DNA sequencing, which allow the determination of the exact sequence of nucleotides or bases in either DNA or RNA. At present it is still a long and expensive process to sequence the entire genome of an organism, and this has been done for only a few species. However, it is quite feasible to determine the sequence of a defined area of a particular chromosome. Typical molecular systematic analyses require the sequencing of around 1000 base pairs.

The Driving Forces of Evolution

Depending on the relative importance assigned to the various forces of evolution, three perspectives provide evolutionary explanations for molecular evolution.

While recognizing the importance of random drift for silent mutations selectionists hypotheses argue that balancing and positive selection are the driving forces of molecular evolution. Those hypotheses are often based on the broader view called panselectionism, the idea that selection is the only force strong enough to explain evolution, relaying random drift and mutations to minor roles.

Neutralists hypotheses emphasize the importance of mutation, purifying selection and random genetic drift The introduction of the neutral theory by Kimura quickly followed by King and Jukes' own findings lead to a fierce debate about the relevance of neodarwinism at the molecular level. The

Neutral theory of molecular evolution states that most mutations are deleterious and quickly removed by natural selection, but of the remaining ones, the vast majority are neutral with respect to fitness while the amount of advantageous mutations is vanishingly small. The fate of neutral mutations are governed by genetic drift, and contribute to both nucleotide polymorphism and fixed differences between species.

Mutationists hypotheses emphasize random drift and biases in mutation patterns Sueoka was the first to propose a modern mutationist view. He proposed that the variation in GC content was not the result of positive selection, but a consequence of the GC mutational pressure.

Chapter–9

DUAL INHERITANCE THEORY

INTRODUCTION

Dual Inheritance Theory (DIT), also known as Gene-Culture Coevolution, was developed in the late 1970s and early 1980s to explain how human behavior is a product of two different and interacting evolutionary processes: genetic evolution and cultural evolution. The DIT is a "middle-ground" between much of social science, which views culture as the primary cause of human behavioral variation, and human sociobiology and evolutionary psychology which view culture as an insignificant by-product of genetic selection. In DIT, culture is defined as information in human brains that got there by social learning. Cultural evolution is considered a Darwinian selection process that acts on cultural information. Dual Inheritance Theorists often describe this by analogy to genetic evolution, which is a Darwinian selection process acting on genetic information Because genetic evolution is relatively well understood, most of DIT examines cultural evolution and the interactions between cultural evolution and genetic evolution.

Theoretical Basis

The DIT holds that genetic and cultural evolution interact in the evolution of Homo sapiens. The DIT recognizes that the natural selection of genotypes is an important component of the evolution of human behavior and that cultural traits can be constrained by genetic imperatives. However, DIT also recognizes that genetic evolution has endowed the human species with a parallel evolutionary process of cultural evolution. DIT makes three main claims:

Culture capacities are adaptations

The human capacity to store and transmit culture arose from genetically evolved psychological mechanisms. This implies that at some point during the evolution of the human species a type of social learning leading to cumulative cultural evolution was evolutionarily advantageous.

Culture evolves

Social learning processes give rise to cultural evolution. Cultural traits are transmitted differently than genetic traits and, therefore, result in different population-level effects. These effects can help explain human behavioral variation.

Genes and culture coevolve

Cultural traits alter the social and physical environments under which genetic selection operates. For example, the cultural adoptions of agriculture and dairying have, in humans, caused genetic selection for the traits to digest starch and lactose, respectively As another example, it is likely that once culture became adaptive, genetic selection caused a refinement of the cognitive architecture that stores and transmits cultural information. This refinement may have further influenced the way culture is stored and the biases that govern its transmission.

The DIT also predicts that, under certain situations, cultural evolution may select for traits that are genetically maladaptive. An example of this is the demographic transition, which describes the fall of birth rates within industrialized societies. Dual Inheritance Theorists hypothesize that the demographic transition may be a result of a prestige bias, where individuals that forgo reproduction to gain more influence in industrial societies are more likely to be chosen as cultural models.

DIT View of Culture

People have defined the word "culture" to describe a large set of different phenomena A definition that sums up what is meant by "culture" in DIT is:

Culture is information stored in individuals' brains that is capable of affecting behaviour and that got there through social learning.

This view of culture emphasizes population thinking by focusing on the process by which culture is generated and maintained. It also views culture as a dynamic property of individuals, as opposed to the more standard social science view of culture as a superorganismic entity to which individuals must conform. This view's main advantage is that it connects individual-level processes to population-level outcomes.

Genetic Influence on Cultural Evolution

Genes have an impact on cultural evolution via psychological predispositions on cultural learning. Genes encode much of the information needed to form the human brain. Genes constrain the brain's structure and, hence, the ability of the brain to store culture. Genes may also endow individuals with certain types of transmission bias.

Cultural Influences on Genetic Evolution

Culture can profoundly influence gene frequencies in a population. One of the best known examples is the prevalence of the genotype for adult lactose absorption in human populations, such as Northern Europeans and some African societies, with a long history of raising cattle for milk. Other societies such as East Asians and Amerindians, retain the typical mammalian genotype in which the body shuts down lactase production shortly after the normal age of weaning. This implies that the cultural practice of raising cattle for milk led to a selection for genetic traits for lactose digestion.

Mechanisms of Cultural Evolution

In DIT, the evolution and maintenance of cultures is described by five major mechanisms, natural selection of cultural variants, random variation, cultural drift, guided variation and transmission bias.

NATURAL SELECTION

Cultural differences among individuals can lead to differential survival of individuals, with those that survive more likely to have their cultural variants adopted by others. The patterns of this selective process depend on transmission biases and can result in behavior that is more adaptive to a given environment.

Random Variation

Random variation arises from errors in the learning, display or recall of cultural information.

Cultural Drift

Cultural drift is a process roughly analogous to genetic drift in evolutionary biology. In cultural drift, the frequency of cultural traits in a population may be subject to random fluctuations due to chance variations in which traits are observed and transmitted (sometimes called "sampling error") These fluctuations might cause cultural variants to disappear from a population. This effect should be especially strong in small populations.

Guided Variation

Cultural traits may be gained in a population through the process of individual learning. Once an individual learns a novel trait, it can be transmitted to other members of the population. The process of guided variation depends on an adaptive standard that determines what cultural variants are learned.

BIASED TRANSMISSION

Understanding the different ways that culture traits can be transmitted between individuals has been an important part of DIT research since the 1970s Transmission biases occur when some cultural variants are favored over others during the process of cultural transmission.

Content Bias

Content biases result from situations where some aspect of a cultural variant's content makes them more likely to be adopted. Content biases can result from genetic preferences, preferences determined by existing cultural traits, or a combination of the two. For example, food preferences can result from genetic preferences for sugary or fatty foods and socially-learned eating practices and taboos. Content biases are sometimes called "direct biases.

Context Bias

Context biases result from individuals using clues about the social structure of their population to determine what cultural variants to adopt. This determination is made without reference to the content of the variant. There are two major categories of context biases: (1) model-based biases; and (2) frequency-dependent biases.

Model-based Biases

Model-based biases result when an individual is biased to choose a particular "cultural model" to imitate. There are four major categories of model-based biases: (1) prestige bias; (2) skill bias; (3) success bias; (4) similarity bias. A "prestige bias" results when individuals are more likely to imitate cultural models that are seen as having more prestige. A measure of prestige could be the amount of deference shown to a potential cultural model by other individuals. A "skill bias" results when individuals can directly observe different cultural models performing a learned skill and are more likely to imitate cultural models that perform better at the specific skill. A "success bias" results from individuals preferentially imitating cultural models that they determine are most generally successful (as opposed to successful at a specific skill as in the skill bias. A "similarity bias" results when individuals are more likely to imitate cultural models that are perceived as being similar to the individual based on specific traits.

Frequency-dependent Biases

Frequency-dependent biases result when an individual is biased to choose particular cultural variants based on their perceived frequency in the population. The most explored frequency-dependent bias is the "conformity bias". Conformity biases result when individuals attempt to copy the mean or the mode cultural variant in the population. Another possible frequency dependent bias is the "rarity bias". The rarity bias results when individuals preferentially choose cultural variants that are less common in the population. The rarity bias is also sometimes called a "non-conformist bias".

Social Learning and Cumulative Cultural Evolution

In DIT, the evolution of culture is dependent on the evolution of social learning. Analytic models show that social learning becomes evolutionarily beneficial when the environment changes with enough frequency that genetic inheritance cannot track the changes, but not fast enough that individual learning is more efficient While other species have social learning, and thus some level of culture, only humans, some birds and chimpanzees are known to have cumulative culture. Researchers argue that the evolution of cumulative culture depends on observational learning and is uncommon in other species because it is ineffective when it is rare in a population. They propose that the environmental changes occurring in the Pleistocene may have provided the right environmental conditions. Researchers argue that cumulative cultural evolution results from a "ratchet effect" that began when humans developed the cognitive architecture to understand others as mental agents. Furthermore it is proposed in the 80s that there are some disparities between the observational learning mechanisms found in humans and great apes - which go some way to explain the observable difference between great ape traditions and human types of culture.

CULTURAL GROUP SELECTION

Although group selection is commonly thought to be nonexistent or unimportant in genetic evolution, DIT predicts that, due to the nature of cultural inheritance, it may be an important force in cultural evolution. The reason group selection is thought to operate in cultural evolution is because of conformist biases. Conformist biases make it difficult for novel cultural traits to spread through a population. Conformist bias also helps maintain variation between groups. These two properties, rare in genetic transmission, are necessary for group selection to operate Based on an earlier model by the conformist biases are almost inevitable when traits spread through social learning, implying that group selection is common in cultural evolution. Group selection can also only work if the rate of group formation is greater than the rate of group extinction. Analysis of the rates of formation and extinction of small groups in New Guinea implied that cultural group selection might be a good explanation for slowly changing aspects of social structure, but not for rapidly changing fads. The ability of cultural evolution to maintain intergroup diversity is what allows for the study of cultural phylogenetics.

Historical Development

The idea that human cultures undergo a similar evolutionary process as genetic evolution goes back at least to Darwin, In the 1960s Donald T. Campbell published some of the first theoretical work that adapted principles of evolutionary theory to the evolution of cultures. In 1976 two developments in cultural evolutionary theory set the stage for DIT. In that year Richard Dawkins's *The Selfish Gene* introduced ideas of cultural evolution to a popular audience. Although one of the best-selling science books of all time, because of its lack of mathematical rigor, it had little impact on the development of DIT. Also in 1976, geneticists Marcus Feldman and Luigi Luca Cavalli-Sforza published the first dynamic models of gene-culture coevolution. These models were to form the basis for subsequent work on DIT, heralded by the publication of three seminal books in 1980 and 1981.

The first was Charles Lumsden and E.O. Wilson's *Genes, Mind and Culture.* This book outlined a series of mathematical models of how genetic evolution might favour the selection of cultural traits and how cultural traits might, in turn, affect the speed of genetic evolution. While it was the first book published describing how genes and culture might coevolve, it had relatively little impact on the further development of DIT Some critics felt that their models depended too heavily on genetic mechanisms at the expense of cultural mechanisms Controversy surrounding Wilson's sociobiological theories may also have decreased the lasting impact of this book.

The second book was Cavalli-Sforza and Feldman's Cultural Transmission and Evolution: A Quantitative Approach (1981). Borrowing heavily from population genetics and epidemiology, this book built a mathematical theory concerning the spread of cultural traits. It describes the evolutionary implications of vertical transmission, passing cultural traits from parents to offspring; oblique transmission, passing cultural traits from any member of an older generation to a younger generation; and horizontal transmission, passing traits between members of the same population.

The next significant DIT publication was Robert Boyd and Peter Richerson's Culture and the Evolutionary Process (1985). This book presents the now-standard mathematical models the evolution of social learning under different environmental conditions, the population effects of social learning, various forces of selection on cultural learning rules, different forms of biased transmission and their population-level effects, and conflicts between cultural and genetic evolution. The book's conclusion also outlined areas for future research that are still relevant today.

Current and Future Research

In their 1985 book, Boyd and Richerson outlined an agenda for future DIT research. This agenda, called for the development of both theoretical models and empirical research. DIT has since built a rich tradition of theoretical models over the past two decades. However, there has not been a comparable level of empirical work.

In a 2006 interview Harvard biologist E.O. Wilson expressed disappointment at the little attention afforded to DIT:

> ". . . for some reason I haven't fully fathomed, this most promising frontier of scientific research has attracted very few people and very little effort."

Kevin Laland and Gillian Brown attribute this lack of attention to DIT's heavy reliance on formal modeling, which doesn't attract the media attention of less rigorous approaches to human behavioral evolution, such as evolutionary psychology:

> "In many ways the most complex and potentially rewarding of all approaches, [DIT], with its multiple processes and cerebral onslaught of sigmas and deltas, may appear too abstract to all but the most enthusiastic reader. Until such a time as the theoretical hieroglyphics can be translated into a respectable empirical science most observers will remain immune to its message."

Economist Herbert Gintis disagrees with this critique, citing existing empirical work as well as more recent work using techniques from behavioral economics. These behavioral economic techniques have been adapted to test predictions of cultural evolutionary models in laboratory settings as well as studying differences in cooperation in fifteen small-scale societies in the field.

Since one of the goals of DIT is to explain the distribution of human cultural traits, ethnographic and ethnologic techniques may also be useful for testing hypothesis stemming from DIT. Although findings from traditional ethnologic studies have been used to buttress DIT arguments, thus far there have been little ethnographic fieldwork designed to explicitly test these hypothese A major difficulty in using existing ethnographic data to test DIT is that, due to cultural anthropology's assumption of culture as a superorganic entity, ethnographic data tends to ignore individual and intragroup cultural variation and focus almost entirely on intergroup variation.

The DIT has been viewed as having great potential for unifying diverse academic fields under one overarching theory. It is identified DIT as the ideal way to build a comprehensive theory of cultural evolution to answer questions about human behavior at different temporal and spacial scales. Along with game theory, Herb Gintis has named DIT one of the two major conceptual theories with potential for unifying the behavioral sciences, including economics, biology, anthropology, sociology, psychology and political science. Because it addresses both the genetic and cultural components of human inheritance, Gintis sees DIT models as providing the best explanations for the ultimate cause of human behavior and the best paradigm for integrating those disciplines with evolutionary theory. In a review of competing evolutionary perspectives on human behavior, Laland and Brown see DIT as the best candidate for uniting the other evolutionary perspectives under one theoretical umbrella.

Relation to Other Fields

Sociology and Cultural Anthropology

Two major topics of study in both sociology and cultural anthropology are human cultures and cultural variation. However, Dual Inheritance theorists charge that both disciplines too often treat culture as a static superorganic entity that dictates human behavior. Cultures are defined by a suite of common traits shared by a large group of people, without regard to variation in cultural traits at the individual level. This is in sharp contrast to DIT, which models human culture at the individual level and views culture as the result of a dynamic evolutionary process at the population level.

Human Sociobiology and Evolutionary Psychology

Human sociobiologists and evolutionary psychologists try to understand how maximizing genetic fitness, in either the modern era or past environments, can explain human behavior. To the human sociobiologist and evolutionary psychologist, culture is either trivial or so bound by genetic fitness that it is unimportant. When faced with a common and seemingly maladaptive trait, practitioners from these disciplines try to determine how the trait actually increases genetic fitness (maybe through kin selection or by speculating about early evolutionary environments). Dual Inheritance theorists, in contrast, will consider a variety of genetic and cultural processes in addition to natural selection on genes.

Human Behavioural Ecology

Human behavioural ecology (HBE) and DIT have a similar relationship to what ecology and evolutionary biology have in the biological sciences. HBE is more concerned about ecological process and DIT more focused on historical process. One difference is that human behavioral ecologists often assume that culture is a system that produces the most adaptive outcome in a given environment. This implies that similar behavioral traditions should be found in similar environments. However, this is not always the case. A study of African cultures showed that cultural history was a better predictor of cultural traits than local ecological conditions

Memetics

Memetics, which comes from the meme idea described in Dawkins's The Selfish Gene, is similar to DIT in that it treats culture as an evolutionary process that is distinct from genetic transmission. However, there are some philosophical differences between memetics and DIT. One difference is that memetics' focus is on the selection potential of the cultural variant (meme) within meme populations, where DIT allows for cultural transmission of non-replicators. Also, DIT does not always view cultural traits as replicators or that replicators are necessary for cumulative adaptive evolution, although in DIT models often assume that cultural variants are replicated (i.e., they model memes). DIT also more strongly emphasizes the role of genetic inheritance in shaping the capacity for cultural evolution. But perhaps the biggest difference is a difference in academic lineage. While Memetics is much better known in popular culture, it is less influential in academia. Authors performing memetic research often do not use the memetic label, perhaps because this can invite controversy. Memetics is often criticized as lacking empirical support or being conceptually ill-founded, and questions are often raised about whether there is hope for the memetic research program succeeding.

BEHAVIOURAL GENETICS

Behavioural genetics is the field of biology that studies the role of genetics in animal (including human) behaviour. The field is an overlap of genetics, ethology and psychology. Classically, behavioural geneticists have studied the inheritance of behavioural traits.

History

In 1869, Francis Galton published the first empirical work in human behavioural genetics, Hereditary Genius. Here, Galton intended to demonstrate that "a man's natural abilities are derived by inheritance, under exactly the same limitations as are the form and physical features of the whole organic world." Like most seminal work, he overstated his conclusions. His was a family study on the inheritance of giftedness and talent. Galton was aware that resemblance among familial relatives can be a function of both shared genes and shared environments. Contemporary behavioural genetics studies special populations—in human research, twins and adoptees and in animal research, specially bred strains and lines—to separate genetic from environmental effects.

The initial impetus behind behavioural genetic research was to demonstrate that there were indeed genetic influences on behaviour. In psychology, this phase lasted for the first half of the 20th century largely because of the overwhelming influence of behaviourism in the field. Later behavioural genetic research focused on quantitative methods, and today there is a large emphasis on applying techniques from molecular genetics to analyse individual genes that influence behaviour.

Contemporary Behavioural Genetics

Currently, the largest branch of human behavioural genetics is psychiatric genetics which studies phenotypes such as schizophrenia, bipolar disorder, and alcoholism.

Recent trends in behaviour genetics have indicated an additional focus toward researching the inheritance of human characteristics typically studied in developmental psychology. For instance, a major focus in developmental psychology has been to characterize the influence of parenting styles on children. However, in most studies, genes are a confounding variable. Because children share half of their genes with each parent, any observed effects of parenting styles could be effects of having many of the same genes as a parent (e.g. harsh aggressive parenting styles have been found to correlate with similar aggressive child characteristics: is it the parenting or the genes?). Thus, behaviour genetics research is currently undertaking to distinguish the effects of the family environment from the effects of genes. This branch of behaviour genetics research is becoming more closely associated with mainstream developmental psychology and the sub-field of developmental psychopathology as it shifts its focus to the heritability of such factors as emotional self-control, attachment, social functioning, aggressiveness, etc.

Several academic bodies exist to support behaviour genetic research, including the International Behavioural and Neural Genetics Society, Behavior Genetics Association, the International Society for Psychiatric Genetics, and the International Society for Twin Studies. Behavior genetic work features prominently in several more general societies, for instance the International Behavioral Neuroscience Society.

Methods of Human Behavioural Genetics

Human behavioural geneticists use several designs to answer questions about the nature and mechanisms of genetic influences on behavior. All of these designs are unified by being based around human relationships which disentangle genetic and environmental relatedness.

So, for instance, some researchers study adopted twins: the adoption study. In this case the adoption disentangles the genetic relatedness of the twins (either 50% or 100%) from their family environments. Likewise the classic twin study contrasts the differences between identical twins and fraternal twins within a family compared to differences observed between families. This core design can be extended: the so-called "extended twin study" which adds additional family members, increasing power and allowing new genetic and environmental relationships to be studied. Excellent examples of this model are the Virginia 20,000 and the QIMR twin studies.

Also possible are the "children of twins" design (holding maternal genetic contributions equal across children with paternal genetics and family environments; and the "virtual twins" design - unrelated children adopted into a family who are very close or identical in age to biological children or other adopted children in the family. While the classical twin study has been criticized they continue to be of high utility. There are several dozen major studies ongoing, in countries as diverse as the USA, UK, Germany, France, The Netherlands, and Australia, and the method is used widely in fields as diverse as dental caries, BMI, aging, substance abuse, sexuality, cognitive abilities, personality, values, and a wide range of psychiatry disorders. This is broad utility is reflected in several thousands of peer-review papers, and several dedicated societies and journals.

Chapter–10

PHENETICS

INTRODUCTION

In biology, phenetics, also known as numerical taxonomy or taximetrics, is an attempt to classify organisms based on overall similarity, usually in morphology or other observable traits, regardless of their phylogeny or evolutionary relation.

Phenetics has largely been superseded by cladistics for research into evolutionary relationships among species. However, certain phenetic methods, such as neighbor-joining, have found their way into cladistics, as a reasonable approximation of phylogeny when more advanced methods (such as Bayesian inference) are too computationally expensive.

Phenetic techniques include various forms of clustering and ordination. These are sophisticated ways of reducing the variation displayed by organisms to a manageable level. In practice this means measuring dozens of variables, and then presenting them as two or three dimensional graphs. Much of the technical challenge in phenetics revolves around balancing the loss of information in such a reduction against the ease of interpreting the resulting graphs.

DIFFERENCE FROM CLADISTICS

Phenetic analyses do not distinguish between plesiomorphies - traits that are inherited from an ancestor (and therefore phylogenetically uninformative) - and apomorphies - traits that evolved anew in one or several lineages. Consequently, phenetic analyses are liable to be misled by convergent evolution and adaptive radiation. A typical error occurring in phenetic analysis is that basal evolutionary grades - which retain many plesiomorphies compared to more advanced lineages - appear to be monophyletic.

Consider for example songbirds. These can be divided into two groups - Corvida, which retains ancient characters in phenotype and genotype, and Passerida, which has more modern traits. But only the latter are a group of closest relatives; the former are numerous independent and ancient lineages which are about as distantly related to each other as each of them is to the more modern songbirds. In a phenetic analysis, the large degree of overall similarity found among the former will make them appear to be monophyletic too, but their shared traits were present in the ancestors of all songbirds. It is the loss of these ancestral traits rather than their presence that signifies which songbirds are more closely related to each other than to other songbirds.

But the two methodologies need not be mutually exclusive. In general, phenetics is today recognized to provide little if any information about the evolutionary relationships among taxa. But there is no reason why e.g. species identified using phenetics cannot subsequently be subjected to cladistic analysis, to determine its evolutionary relationships.

Phenetic methods can be superior to cladistics when only the distinctness of related taxa is important, as the computational requirements are lower. On the other hand, whenever information on the evolutionary history of taxa is needed for a study, cladistic methods are used today.

Phenetics Today

Traditionally there was a great deal of heated debate between pheneticists and cladists, as both methods were initially proposed to resolve evolutionary relationships. Perhaps the "high-water mark" of phenetics were the DNA-DNA hybridization studies by Charles G. Sibley, Jon E. Ahlquist and Burt L. Monroe Jr., from which resulted the 1990 Sibley-Ahlquist taxonomy for birds. Highly controversial at its time, some of its findings (e.g. the Galloanserae) have been vindicated, while others (e.g. the all-inclusive "Ciconiiformes" or the "Corvida") have been rejected. However, with computers growing increasing powerful and widespread, more refined cladistic algorithms became available and could put the suggestions of Willi Hennig to the test; as it turned out, the results of cladistic analyses turned out to be superior to those of phenetic methods - at least when it came to resolving phylogenies.

Many systematists continue to use phenetic methods, particularly in addressing species-level questions. While the ultimate goal of taxonomy includes finding the 'tree of life' - the evolutionary path connecting all species - in fieldwork one needs to be able to separate one taxon from another. Classifying diverse groups of closely-related organisms that differ by very subtle differences is difficult using a cladistic approach. Phenetics provides numerical tools for examining overall patterns of variation, allowing researchers to identify discrete groups that can be classified as species.

Modern applications of phenetics are common in botany, and some examples can be found in most issues of the journal Systematic Botany. Indeed, due to the effects of horizontal gene transfer, polyploid complexes and other peculiarities of plant genomics, phenetic techniques in botany - though less informative altogether - are also less prone to errors compared cladistic analysis of DNA sequences.

In addition, many of the techniques developed by phenetic taxonomists have been adopted and extended by community ecologists, due to a similar need to deal with large amounts of data.

CLADE

A clade is a monophyletic group - that is, a single common ancestor and all its descendants. The common ancestor of any reasonably sized group, and many of its descendants, will be not be living It is not necessary for a clade to contain living representatives.

Cladistics and Linnean taxonomy have an uneasy relationship. Because Linnean taxonomy requires that nature is split into nameable pigeon-holes (species, genera, and higher taxa), it breaks down over evolutionary time, because species arise by gradual modification, not step changes. There is no biological basis to make a distinction between one species and a 'descendant' species. This is where cladistics trumps Linnean taxonomy; intermediate taxa can be named according to their relationship to named taxa using the stem group terminology ; the disadvantage, however, is that sibling or parent-descendant pairs belong to different clades despite being, for all practical purposes, identical organisms.

Naming Clades

Valid taxa are monophyletic groups - that is, they are clades. However, traditional Linnaean taxonomy does not cope well in a cladistic framework: for instance, it cannot adequately classify stem groups.

Three methods of naming clades have been proposed: node-, stem-, and apomorphy-based. In node-based naming, taxon name A might refer to the least inclusive clade containing X and Y. In stem-based naming, A would refer to the most inclusive clade containing X and Y but not Z. In apomorphy (derived feature)-based naming, A would refer to the clade identified by a feature synapomorphic (sharing a derivation) with a feature in specimen (taxon) X. Differences between a traditional approach and these phylogenetic alternatives become obvious when the phylogenetic hypothesis changes.

Comparison between the traditional Linnaean approach to nomenclature and a phylogenetic alternative (node-based naming). Suppose that all we want to do is to give a name ("A") to a clade containing X and Y. In the Linnaean system this means that we have to introduce names for sister taxa, assigning all taxa to the categories species, genus, and family, and designate type species. No explicit reference to phylogeny is made. The phylogenetic alternative provides an explicit reference to evolutionary history, and nothing but the clade containing X and Y needs to be named. When the hypothesis of relationship changes, the phylogenetic alternative is cleaner and more explicit about what it refers to.

CLADISTICS

Cladistics is the hierarchical classification of species based on evolutionary ancestry. Cladistics is distinguished from other taxonomic systems because it focuses on evolution rather than similarities between species, and because it places heavy emphasis on objective, quantitative analysis. Cladistics generates diagrams called cladograms that represent the evolutionary tree of life. DNA and RNA sequencing data are used in many important cladistic efforts. Computer programs are widely used in cladistics, due to the highly complex nature of cladogram generation procedures. Cladistics originated in the work of the German entomologist, Willi Hennig, who himself referred to it as phylogenetic systematics; the use of the terms "cladistics" and "clade" was popularized by other researchers. The term phylogenetics is often used synonymously with cladistics. Cladistics originated in the field of biology but in recent years has found application in other disciplines. The word cladistics is derived from the ancient , klados, "branch."

Terminology

The yellow group (sauropsids) is monophyletic, the blue group (reptiles) is paraphyletic, and the red group (warm-blooded animals) is polyphyletic.

A clade is an ancestor and all of its descendents.

- A monophyletic group is a clade.
- A paraphyletic group is a monophyletic group that excludes some of the descendants (e.g. reptiles are sauropsids excluding birds). Most cladists discourage the use of paraphyletic groups.

- A polyphyletic group is a group consisting of members from two non-overlapping monophyletic groups (e.g. flying animals). Most cladists discourage the use of polyphyletic groups.
- An outgroup is an organism that is considered not to be part of the group in question, but is closely related to the group.
- A characteristic that is present in both the outgroups and in the ancestors is called a plesiomorphy (meaning "close form", also called an ancestral state).
- A characteristic that occurs only in later descendants is called an apomorphy (meaning "separate form", also called a "derived" state) for that group. Note: The adjectives plesiomorphic and apomorphic are used instead of "primitive" and "advanced" to avoid placing value-judgments on the evolution of the character states, since both may be advantageous in different circumstances. It is not uncommon to refer informally to a collective set of plesiomorphies as a ground plan for the clade or clades they refer to.
- A species or clade is basal to another clade if it holds more plesiomorphic characters than that other clade. Usually a basal group is very species-poor as compared to a more derived group. It is not a requirement that a basal group be extant. For example, palaeodicots are basal to flowering plants.
- A clade or species located within another clade is said to be nested within that clade.

Three Definitions of Clade

There are three ways to define a clade for use in a cladistic taxonomy.

- *Node-based:* the most recent common ancestor of A and B, and all its descendants. See crown group.
- *Stem-based:* all descendants of the oldest common ancestor of A and B that is not also an ancestor of Z. See total group.

- *Apomorphy-based:* the most recent common ancestor of A and B, along with all of its descendants, possessing a certain derived character. This definition is generally discouraged by most cladists.

History of Cladistics

Hennig's major book, even the 1979 version, does not contain the term cladistics in the index. He referred to his own approach as phylogenetic systematics, as implied by the book's title. A review paper by Dupuis observes that the term clade was introduced in 1958 by Julian Huxley, cladistic by Cain and Harrison in 1960, and cladist (for an adherent of Hennig's school) by Mayr in 1965.

From the time of Hennig's original formulation until the end of the 1980s cladistics remained a minority approach to classification. However in the 1990s it rapidly became the dominant method of classification in evolutionary biology. Cheap but increasingly powerful personal computers made it possible to process large quantities of data about organisms and their characteristics. At about the same time the development of effective polymerase chain reaction techniques made it possible to apply cladistic methods of analysis to biochemical features of organisms as well as to anatomical ones.

Cladistics as a Successor to Phenetics

For some decades in the mid to late 20th century, a commonly used methodology was phenetics ("numerical taxonomy"). This can be seen as a predecesso to some methods of today's cladistics (namely distance matrix methods like neighbor-joining), but made no attempt to resolve phylogeny, only similarities.

Cladograms

The starting point of cladistic analysis is a group of species and molecular, morphological, or other data characterizing those species. The end result is a tree-like relationship diagram called a cladogram, or sometimes a dendrogram (Greek for "tree drawing").The cladogram graphically represents a hypothetical evolutionary process. Cladograms are subject to revision as additional data become available.

Synonyms

The terms evolutionary tree, and sometimes phylogenetic tree are often used synonymously with cladogram,but others treat phylogenetic tree as a broader term that includes trees generated with a nonevolutionary emphasis.

Subtrees are Clades

In cladograms, all organisms lie at the leaves. The two taxa on either side of a split are called sister taxa or sister groups. Each subtree, whether it contains only two or a hundred thousand items, is called a clade.

2-way Versus 3-way Forks

Many cladists require that all forks in a cladogram be 2-way forks. Some cladograms include 3-way or 4-way forks when there are insufficient data to resolve the forking to a higher level of detail. See phylogenetic tree for more information about forking choices in trees.

Depth

If a cladogram represents N species, the number of levels (the "depth") in the cladogram is on the order of $\log^2(N)$. For example, if there are 32 species of deer, a cladogram representing deer will be around 5 levels deep (because 25 = 32). A cladogram representing the complete tree of life, with about 10 million species, would be about 23 levels deep. This formula gives a lower limit: in most cases the actual depth will be a larger value because the various branches of the cladogram will not be uniformly deep. Conversely, the depth may be shallower if forks larger than 2-way forks are permitted.

Time Scale

A cladogram tree has an implicit time axis with time running forward from the base of the tree to the leaves of the tree. If the approximate date (for example, expressed as millions of years ago) of all the evolutionary forks were known, those dates could be captured in the cladogram. Thus, the time axis of the cladogram could be assigned a time scale (e.g. 1 cm = 1 million years), and the forks of the tree could be graphically located along the time axis. Such cladograms are called scaled cladograms. Many cladograms are not scaled along the time axis, for a variety of reasons.

Many cladograms are built from species characteristics that cannot be readily dated (e.g. morphological data in the absence of fossils or other dating information)

When the characteristic data are DNA/RNA sequences, it is feasible to use sequence differences to establish the relative ages of the forks, but converting those ages into actual years requires a significant approximation of the rate of change.

Even when the dating information is available, positioning the cladogram's forks along the time axis in proportion to their dates may cause the cladogram to become difficult to understand or hard to fit within a human-readable format.

Extinct Species

Cladistics makes no distinction between extinct and nonextinct species and it is appropriate to include extinct species in the group of organisms being analyzed. Cladograms that are based on DNA/RNA generally do not include extinct species because DNA/RNA samples from extinct species are rare. Cladograms based on morphology, especially morphological characteristics that are preserved in fossils, are more likely to include extinct species.

Cladistics Contrasted with Traditional Taxonomy

A highly resolved, automatically generated tree of life based on completely sequenced genomes. Prior to the advent of cladistics, most taxonomists used Linnaean taxonomy and later Evolutionary taxonomy to organize life forms. These traditional approaches, still in use by some researchers (especially in works intended for a more general audience use several fixed levels of a hierarchy, such as kingdom, phylum, class, order, and family. Cladistics does not use those terms, because one of the fundamental premises of cladistics is that the evolutionary tree is so deep and so complex that it is inadvisable to set a fixed number of levels.

Evolutionary taxonomy insists that groups reflect phylogenies. In contrast, Linnean taxonomy allows both monophyletic and paraphyletic groups as taxa. Since the early 20th century, Linnaean taxonomists have generally attempted to make genus-level and lower-level taxa monophyletic. Ernst Mayr drew a distinction between the terms cladistics and phylogeny, using the term cladistics to refer to classifications which only take into account genealogy, as opposed to phylogeny, which had previously been used in a broader sense to refer to the combination of genealogy and amount of divergence from an ancestor (i.e. Evolutionary taxonomy). Mayr wrote, in 1985:

> It would seem to me to be quite evident that the two concepts of phylogeny (and their role in the construction of classifications) are sufficiently different to require terminological distinction. The term phylogeny should be retained for the broad concept of phylogeny, promoted by Darwin and adopted by most students of phylogeny in the ensuing 90 years. The concept of phylogeny as mere genealogy should be terminologically distinguished as cladistics. To lump the two concepts together terminologically could not help but produce harmful equivocation.

Willi Hennig's pioneering work provoked a spirited debate about the relative merits of cladistics versus traditional taxonomy which has continued down to the present Some of the debates that the cladists engaged in had been running since the 19th century, but they entered these debates with a new fervor, as can be seen from the Foreword to Hennig by Rosen, Nelson, and Patterson.

Encumbered with vague and slippery ideas about adaptation, fitness, biological species and natural selection, neo-Darwinism (summed up in the "evolutionary" systematics of Mayr and Simpson) not only lacked a definable investigatory method, but came to depend, both for evolutionary interpretation and classification, on consensus or authority.

Cladistics strictly and exclusively follows phylogeny and has arbitrarily deep trees with binary branching: each taxon is a clade. Linnaean taxonomy, while following phylogeny, also subjectively considers morphology and has a fixed hierarchy, whose taxa are not always clades.

Paraphyletic Groups Discouraged

Many cladists discourage the use of paraphyletic groups because they detract from cladistics' emphasis on clades (monophyletic groups). In contrast, proponents of the use of paraphyletic groups argue that any dividing line in a cladogram creates both a monophyletic section above and a paraphyletic section below. They also contend that paraphyletic taxa are necessary for classifying earlier sections of the tree – for instance, the early vertebrates that would someday evolve into the family Hominidae cannot be placed in any other monophyletic family. They also argue that paraphyletic taxa provide information about significant changes in organisms' morphology, ecology, or life history – in short, that both paraphyletic groups and clades are valuable notions with separate purposes.

Complexity of the Tree of Life

One argument in favor of cladistics is that it supports arbitrarily complex, arbitrarily deep trees. Especially when extinct species are considered (both known and unknown), the complexity and depth of the tree can be very large. Every single speciation event, including all the species that are now extinct, represents an additional fork on the hypothetical, complete cladogram representing the full tree of life. Fractals can be used to represent this notion of increasing detail: as a viewpoint zooms into the tree of life, the complexity remains virtually constant . This great complexity of the tree, and the uncertainty associated with the complexity, are among the reasons that cladists cite for the attractiveness of cladistics over traditional taxonomy.

Proponents of noncladistic approaches to taxonomy point to punctuated equilibrium to bolster the case that the tree of life has a finite depth and finite complexity. If the number of species currently alive is finite, and the number of extinct species that we will ever know about is finite, then the depth and complexity of the tree of life is bounded, and there is no need to handle arbitrarily deep trees.

Phylo Code Approach to Naming Species

A formal code of phylogenetic nomenclature, the PhyloCode is currently under development for cladistic taxonomy. It is intended for use by both those who would like to abandon Linnaean taxonomy and those who would like to use taxa and clades side by side. In several instances (see for example Hesperornithes) it has been employed to clarify uncertainties in Linnaean systematics so that in combination they yield a taxonomy that unambiguously places problematic groups in the evolutionary tree in a way that is consistent with current knowledge.

Example

For example, Linnaean taxonomy contains the taxon Tetrapoda, defined morphologically as vertebrates with four limbs (as well as animals with four-limbed ancestors, such as snakes), which is often given the rank of superclass, and divides into the classes Amphibia, Reptilia, Aves, Mammalia, and some extinct families.

Cladistics also contains the taxon Tetrapoda, whose living members can be classified phylogenically as "the clade defined by the common ancestor of amphibians and mammals", or more precisely the clade defined by the common ancestor of a specific amphibian and mammal (or bird or reptile), but whose tree is still being worked out (there are a number of extinct branches). The taxon does not have a rank, and its subtaxa are subclades:

these can be contained within one another, but one does not divide the clade into several non-overlapping taxa (as in traditional taxonomy): one can split into two clades at the first branching, but that is all. With regards to the traditional classes, Aves and Mammalia are subclades, contained in the subclade Amniota, but Reptilia is a paraphyletic taxon, not a clade — "At best, the cladists suggest, we could say that the traditional Reptilia are "non-avian, non-mammalian amniotes"— and instead one divides Amniota into the two clades Sauropsida (which contains birds and all living amniotes other than mammals, including all living traditional reptiles) and Theropsida (mammals and the extinct "mammal-like reptiles"). Similarly, Amphibia is a paraphyletic taxon.

The characteristics used to create a cladogram can be roughly categorized as either morphological (synapsid skull, warm blooded, notochord, unicellular, etc.) or molecular (DNA, RNA, or other genetic information) Prior to the advent of DNA sequencing, all cladistic analysis used morphological data.

As DNA sequencing has become cheaper and easier, molecular systematics has become a more and more popular way to reconstruct phylogenies. Using a parsimony criterion is only one of several methods to infer a phylogeny from molecular data; maximum likelihood and Bayesian inference, which incorporate explicit models of sequence evolution, are non-Hennigian ways to evaluate sequence data. Another powerful method of reconstructing phylogenies is the use of genomic retrotransposon markers, which are thought to be less prone to the problem of reversion that plagues sequence data. They are also generally assumed to have a low incidence of homoplasies because it was once thought that their integration into the genome was entirely random; this seems at least sometimes not to be the case, however.

Ideally, morphological, molecular, and possibly other phylogenies should be combined into an analysis of total evidence: All have different intrinsic sources of error. For example, character convergence (homoplasy) is much more common in morphological data than in molecular sequence data, but character reversions that are unrecognizable as such are more common in the latter (see long branch attraction). Morphological homoplasies can usually be recognized as such if character states are defined with enough attention to detail.

PLESIOMORPHIES AND SYNAPOMORPHIES

The researcher must decide which character states were present before the last common ancestor of the species group (plesiomorphies) and which were present in the last common ancestor (synapomorphies) and does so by comparison to one or more outgroups. The choice of an outgroup is a crucial step in cladistic analysis because different outgroups can produce trees with profoundly different topologies. Note that only synapomorphies are of use in characterizing clades.

Avoid Homoplasies

A homoplasy is a character that is shared by multiple species due to some cause other than common ancestry. Typically, homoplasies occur due to convergent evolution. Use of homoplasies when building a cladogram is sometimes unavoidable but is to be avoided when possible.

A well known example of homoplasy due to convergent evolution would be the character, "presence of wings". Though the wings of birds, bats, and insects serve the same function, each evolved independently, as can be seen by their anatomy. If a bird, bat, and a winged insect were scored for the character, "presence of wings", a homoplasy would be introduced into the dataset, and this would confound the analysis, possibly resulting in a false evolutionary scenario.

Homoplasies can often be avoided outright in morphological datasets by defining characters more precisely and increasing their number. When analyzing "supertrees" (datasets incorporating as many taxa of a suspected clade as possible), it may become unavoidable to introduce character definitions that are imprecise, as otherwise the characters might not apply at all to a large number of taxa; to continue with the "wings" example, the presence of wings would hardly be a useful character if attempting a phylogeny of all Metazoa, as most of these don't have wings at all. Cautious choice and definition of characters thus is another important element in cladistic analyses. With a faulty outgroup or character set, no method of evaluation is likely to produce a phylogeny representing the evolutionary reality.

Application to Other Disciplines

A triple family tree of Linux distributions.The processes used to generate cladograms are not limited to the field of biology The generic nature of cladistics means that cladistics can be used to organize groups of items in many different academic realms. The only requirement is that the items have characteristics that can be identified and measured.

Chapter–11

Phylogenetic Comparative Methods

INTRODUCTION

Phylogenetic comparative methods (PCMs) use information on the evolutionary relationships of organisms (phylogenetic trees) to compare species. The most common applications are to test for correlated evolutionary changes in two or more traits, or to determine whether a trait contains phylogenetic signal (the tendency for related species to resemble each other). However, several methods are available to relate particular phenotypic traits to variation in rates of speciation and/or extinction, including attempts to identify evolutionary key innovations. Although most studies that employ PCMs focus on extant organisms, the methods can also be applied to extinct taxa and can incorporate information from the fossil record.

Owing to their computational requirements, they are usually implemented by computer programs. PCMs can be viewed as part of evolutionary biology, systematics, phylogenetics, bioinformatics or even statistics, as most methods involve statistical procedures and principles for estimation of various parameters and drawing inferences about evolutionary processes.

What distinguishes PCMs from most traditional approaches in systematics and phylogenetics is that they typically do not attempt to infer the phylogenetic relationships of the species under study. Rather, they use an independent estimate of the phylogenetic tree (topology plus branch lengths) that is derived from a separate phylogenetic analysis, such as comparative DNA sequences that have been analyzed by maximum parsimony or maximum likelihood methods. PCMs are consumers of phylogenetic trees, not primary producers of them. Accordingly, the list of phylogenetics software shows little overlap with the programs for PCMs.

Many other fields of biology use interspecific comparison as well, including behavioural ecology, ethology, ecophysiology, comparative physiology, evolutionary physiology, functional morphology, comparative biomechanics, and the study of sexual selection.

The comparative method is also used heavily in linguistics.

Applications

The PCMs can be used to analyze the origin and maintenance of biodiversity. Biodiversity is most commonly discussed in terms of the number of species, but it can also be phrased in terms of the amount of phenotypic (e.g., physiological, morphological) space that a given set of species occupies (see also Cambrian explosion).

Home range areas of 49 species of mammals in relation to their body size. Larger-bodied species tend to have larger home ranges, but at any given body size members of the order Carnivora (carnivores and omnivores) tend to have larger horme ranges than ungulates (all of which are herbivores). Whether this difference is considered statistically significant depends on what type of analysis is applied.

Testes mass of various species of Primates in relation to their body size and mating system. Larger-bodied species tend to have larger testes, but at any given body size species in which females tend to mate with multiple males have males with larger testes.Phylogenetic comparative methods are commonly applied to such questions as:

What is the slope of an allometric scaling relationship?

- *Example:* how does brain mass vary in relation to body mass?

Do different clades of organisms differ with respect to some phenotypic trait?

- *Example:* do canids have larger hearts than felids?

Do groups of species that share a behavioral or ecological feature (e.g., social system, diet) differ in average phenotype?

- *Example:* do carnivores have larger home ranges than herbivores?

What was the ancestral state of a trait?

- *Example:* where did endothermy evolve in the lineage that led to mammals?

Does a trait exhibit significant phylogenetic signal in a particular group of organisms? Do certain types of traits tend to "follow phylogeny" more than others?

- *Example:* are behavioral traits more labile during evolution?

Do species differences in life history traits trade-off, as in the so-called fast-slow continuun?

- *Example:* why do small-bodied species have shorter life spans than their larger relatives?

PHYLOGENETICALLY INDEPENDENT CONTRASTS

The standardized contrasts are used in conventional statistical procedures, with the constraint that all regressions, correlations, analysis of covariance, etc., must pass through the origin.Felsenstein proposed the first general statistical method for incorporating phylogenetic information, i.e., the first that could use any arbitrary topology (branching order) and a specified set of branch lengths. The method is now recognized as an algorithm that implements a special case of what are termed phylogenetic generalized least-squares models.

The logic of the method is to use phylogenetic information (and an assumed Brownian-motion like model of trait evolution) to transform the original tip data (mean values for a set of species) into values that are statistically independent and identically distributed.

The algorithm involves computing values at internal nodes as an intermediate step, but they are generally not used for inferences by themselves. An exception occurs for the basal (root) node, which can be interpreted as an estimate of the ancestral value for the entire tree (assuming that no directional evolutionary trends [e.g., Cope's rule] have occurred) or as a phylogenetically weighted estimate of the mean for the entire set of tip species (terminal taxa). The value at the root is equivalent to that obtained from the "squared-change parsimony" algorithm and is also the maximum likelihood estimate under Brownian motion. The independent contrasts algebra can also be used to compute a standard error or confidence interval.

Phylogenetically Informed Monte Carlo Computer Simulations

It is proposed that one way to account for phylogenetic relations when conducting statistical analyses was to use computer simulations to create many data sets that are consistent with the null hypothesis under test (e.g., no correlation between two traits, no difference between two ecologically defined groups of species) but that mimic evolution along the relevant phylogenetic tree. If such data sets (typically 1,000 or more) are analyzed with the same statistical procedure that is used to analyze a real data set, then results for the simulated data sets can be used to create phylogenetically correct null distributions of the test statistic (e.g., a correlation coefficient, t, F). Such simulation approaches can also be combined with methods like phylogenetically independent contrasts.

Some Interesting Quotations

In addition to having a long and extremely productive history in biology, comparative methods are often controversial.

Rather than attempt to recount the various controversies, many of which are ongoing, quotes (in chronological order) are illustrative.

> "Ought we, for instance, to begin by discussing each separate species - man, lion, ox, and the like - taking each kind in hand independently of the rest, or ought we rather to deal first with the attributes which they have in common in virtue of some common element of their nature, and proceed from this as a basis for the consideration of the separately?" (Aristotle, *De partibus animalium*; quoted in Harvey and Pagel, 1991, p. 35)

> "In parallel with laboratory experimental methods, the comparative method increases in value with its sample size, i.e., the number of species being compared. When species are few and their phylogenetic relationships are clouded in the distant past of unfossilized ancestors, the comparative method reaps fewer conclusions of trust." (Hailman, 1976, p. 20)

> ". . . there is no easy way, except by comparison, to test most questions about the long-term history of life, or to generate predictions from evolutionary considerations." (Alexander, 1979, p. 13)

> ". . . we must learn to treat comparative data with the same respect as we would treat experimental results ..." (Maynard Smith and Holliday, 1979, p. vii)

> "In he past, however, cooperation between systematists and other comparative biologists has been sporadic at best. Most experimental biologists have ignored taxonomy and systematics, some even to the extent of not bothering to provide their animals with proper identifications or scientific names."

> "If we assume that the . . . cladogram . . . is correct, we can then hypothesize what the particular common ancestor must have been like." (Atz et al., 1980, p. 14)

> ". . . biology will never secure fossils of all species in the past because fossilization is such a rare process, requiring

just the right physical and chemical conditions. Therefore, in order to trace probable phylogenetic lineages one must reason from the evidence at hand: the characteristics of contemporary animals themselves which are the end-points of phylogeny (evolutionary history)."

". . . the comparative method of 1950 was indistinguishable from the comparative method of 350 BC . . ." (Ridley, 1983, p. 6)

"Focusing only on highly adapted species may, of course, bring valuable information on extreme situations but might also obscure the basic mechanisms." (Bankir and Rouffignac, 1985, p. R663)

"Most of what we know is based upon comparison. When asked to describe a food not previously tasted or a new kind of music, one often responds that the taste is "like" some other food, or that the sensation "differs" in a particular way from something that is familiar. Indeed, comparison and the similarities and differences it discloses is ingrained in our approach to description of objects, events and processes. Hence the questions "what can we compare?" and its ancillary "how shall we compare?" prove to be the key to any study of natural phenomena."

"Some reviewers of this paper felt that the message was "rather nihilistic," and suggested that it would be much improved if I could present a simple and robust method that obviated the need to have an accurate knowledge of phylogeny. I entirely sympathize, but do not have a method that solves the problem. The best we can do is perhaps to use pairs of close relatives as suggested above, although this discards at least half of the data. Comparative biologists may understandably feel frustrated upon being told that they need to know the phylogenies of their groups in great detail, when this is not something they had much interest in knowing. Nevertheless, efforts to cope with the effects of the phylogeny will have to be made. Phylogenies are fundamental to comparative biology; there is no doing it without taking them into account."

"Comparative biologists tend to suspect comparisons of distantly related species; they hope to base their comparisons on recent evolutionary events that have not been overlaid by much subsequent change."

"Yet, one of the most embarrassing things that could be done to a group of respected biologists would be to ask them to spend a few minutes to write down what is meant by the comparative method, and what are the basic goals and principles of biological comparison."

"As welcome as it is to see the lung successfully employed as a systematic index, it is, on the other hand, unfortunate. Thereby, lungs lose their innocence in the sense that phylogenetic trees and cladograms can no longer be used to help resolve the sequence in the development of lung structure without the danger of circular argumentation."

"To be maximally informative, such studies should be undertaken on closely related groups of organisms, so that factors extraneous to the comparison can be minimized ..."

"The comparative approach is not new. Indeed, it was Darwin's favoured technique. ... In short, comparative studies have taught us most of what we know about adaptation." (Harvey and Pagel, 1991, p. v)

"In short, all useful comparative methods are based on explicit models of evolutionary change." (Harvey and Pagel, 1991, p. v)

"Adaptation is an inherently comparative idea.

"The usual symptom of non-independence is that closely related species tend to be more alike than more distantly related species."

". . . to use species as independent data points in a comparative analysis requires that one ignores phylogenetic relationships."

"Because life-history traits are likely to be correlated with a species' phylogenetic history, unequivocal evidence for adaptation to local environmental conditions may be recognized only after the variation in a trait attributable to phylogeny is removed." (Miles and Dunham, 1992, p. 848)

"Broad-scale comparative evidence from across a large number of taxa may often help set limits to the applicability of hypotheses that have been generated from a particular phenomenon in a particular species."

"Any comparative study lacking a phylogenetic perspective now would be incomplete." (Miles and Dunham, 1993, p. 590)

"However, in general, the evolutionary process involves descent with modification, and in the absence of modification, one must conclude that similarities among closely related taxa reflect shared ancestry. Phylogeny, then, is an important explanatory principle for understanding shared characteristics and should be the null hypothesis in all tests of similarity or differentiation among taxa.")

"Moreover, comparative studies supply only correlational data. Correlation does not demonstrate causation ..." (Garland and Adolph, 1994, p. 823)

"In addition to his theory of natural selection, the comparative method is what made Darwin great. If you don't believe this claim, look at any of his major works. They are packed full with interspecific comparisons based on detailed studies and anecdotal observations."

"Population comparisons can provide particularly powerful means of evaluating adaptive hypotheses for two reasons. The first is that there tend to be fewer differences between populations than between species. Consequently, there are fewer covarying traits to confound analyses . . . Second, divergent populations are often relatively young and may be more likely to reside in the habitats in which their derived character states evolved than is the case for divergent species with potentially longer intervening histories . . . " "Naive, prephylogenetic comparative tests should be kept at the other end of a barge pole." (Ridley and Grafen, 1996, p. 87)

"It is the study of the bizarre, the outliers, the freaks, that gives us some of our clearest insights into the hows and whys of evolution."

Chapter–12

GENETIC DIVERSITY

INTRODUCTION

Genetic diversity is a level of biodiversity that refers to the total number of genetic characteristics in the genetic makeup of a species. It is distinguished from genetic variability, which describes the tendency of genetic characteristics to vary.

The academic field of population genetics includes several hypotheses regarding genetic diversity. The neutral theory of evolution proposes that diversity is the result of the accumulation of neutral substitutions. Diversifying selection is the hypothesis that two subpopulations of a species live in different environments that select for different alleles at a particular locus. This may occur, for instance, if a species has a large range relative to the mobility of individuals within it. Frequency-dependent selection is the hypothesis that as alleles become more common, they become less fit. This is often invoked in host-pathogen interactions, where a high frequency of a defensive allele among the host means that it is more likely that a pathogen will spread if it is able to overcome that allele.

IMPORTANCE OF GENETIC DIVERSITY

There are many different ways to measure genetic diversity. The modern causes for the loss of animal genetic diversity have also been studied and identified. A 2007 study conducted by the National Science Foundation found that genetic diversity and biodiversity are dependent upon each other — that diversity within a species is necessary to maintain diversity among species, and vice versa. According to the lead researcher in the study, Dr. Richard Lankau, "If any one type is removed from the system, the cycle can break down, and the community becomes dominated by a single species."

Survival and Adaptation

Genetic diversity plays a huge role in survival and adaptability of a species. When a species's environment changes, slight gene variations are necessary for it to adapt and survive. A species that has a large degree of genetic diversity among its population will have more variations from which to choose the most fit alleles. Species that have very little genetic variation are at a great risk. With very little gene variation within the species, healthy reproduction becomes increasingly difficult, and offspring often deal with similar problems to those of inbreeding.

When humans initially started farming, they used selective breeding to pass on desirable traits of the crops while omitting the undesirable ones. Selective breeding leads to monocultures: entire farms of nearly genetically identical plants. Little to no genetic diversity makes crops extremely susceptible to widespread disease. Bacteria morph and change constantly. When a disease causing bacteria changes to attack a specific genetic variation, it can easily wipe out vast quantities of the species. If the genetic variation that the bacterium is best at attacking happens to be that which humans have selectively bred to use for harvest, the entire crop will be wiped out

A very similar occurrence is the cause of the infamous Potato Famine in Ireland. Since new potato plants do not come as a result of reproduction but rather from pieces of the parent plant, no genetic diversity is developed, and the entire crop is essentially a clone of one potato, it is especially susceptible to an epidemic. In the 1840s, much of Ireland's population depended on potatoes for food. They planted namely the "lumper" variety of potato, which was susceptible to a rot-causing mold called Phytophthora infestans This mold destroyed the vast majority of the potato crop, and left tens of thousands of people to starve to death.

Coping with poor genetic diversityThe natural world has several ways of preserving or increasing genetic diversity. Among oceanic plankton, viruses aid in the genetic shifting process. Ocean viruses, which infect the plankton, carry genes of other organisms in addition their own. When a virus containing the genes of one cell infects another, the genetic makeup of the latter changes. This constant shift of genetic make-up helps to maintain a healthy population of plankton despite complex and unpredictable environmental changes.

Cheetahs are a threatened species. Extremely low genetic diversity and resulting poor sperm quality has made breeding and survivorship difficult for cheetahs -- only about 5% of cheetahs make survive to adulthood. About 10,000 years ago, all but the jubatus species of cheetahs died out. The species encountered a population bottleneck and close family relatives were forced to mate with each other, or inbreed. However, it has been recently discovered that female cheetahs can mate with more than one male per litter of cubs. They undergo induced ovulation, which means that a new egg is produced every time a female mates. By mating with multiple males, the mother increases the genetic diversity within a single litter of cubs.

GENETIC VARIABILITY

Genetic variability is a measure of the tendency of individual genotypes in a population to vary from one another. Variability is different from genetic diversity, which is the amount of variation seen in a particular population The variability of a trait describes how much that trait tends to vary in response to environmental and genetic influences. Genetic variability in a population is important for biodiversity, because without variability, it becomes difficult for a population to adapt to environmental changes and therefore makes it more prone to extinction. Variability is an important factor in evolution as it affects an individual's response to environmental stress and thus can lead to differential survival of organisms within a population due to natural selection of the most fit variants. Genetic variability also underlies the differential susceptibility of organisms to diseases and sensitivity to toxins or drugs — a fact that has driven increased interest in personalized medicine given the rise of the human genome project and efforts to map the extent of human genetic variation such as the HapMap project.

CAUSES OF VARIABILITY

There are many sources of genetic variability in a population:

Genetic recombination is one source of variability; during meiosis in sexual creatures two homologous chromosomes from the male and the female cross over one another and exchange gene sequences. The chromosomes then split apart and are ready to form an offspring. The cross-over is random and is governed by its own set of genes that code for where crossovers can occur (in cis) and for the mechanism behind the exchange of DNA chunks (in trans). Being controlled by genes means that recombination is also variable in frequency, location, thus it can be selected to increase fitness by nature, because the more recombination the more variability and the more variability the easier it is for the population to handle changes

Immigration, emigration, and translocation – each of these is the movement of an individual into or out of a population. When an individual comes from a previously genetically isolated population into a new one it will increase the genetic variability of the next generation if it reproduces.

Polyploidy – having more than two homologous chromosomes allows for even more recombination during meiosis allowing for even more genetic variability in one's offspring.

Diffuse centromeres – in asexual organisms where the offspring is an exact genetic copy of the parent, there are limited sources of genetic variability. One thing that increased variability, however, is having diffused instead of localized centromeres. Being diffused allows the chromatids to split apart in many different ways allowing for chromosome fragmentation and polyploidy creating more variability.

Genetic mutations – contribute to the genetic variability within a population and can have positive, negative, or neutral effects on an fitness This variability can be easily propagated throughout a population by natural selection if the mutation increases the affected individual's fitness and its effects will be minimized/hidden if the mutation is deleterious. However, the smaller a population and its genetic variability are, the more likely the recessive/hidden deleterious mutations will show up causing genetic drift.

Chapter–13

CONVERGENT EVOLUTION

INTRODUCTION

Convergent evolution describes the acquisition of the same biological trait in unrelated lineages. The wing is a classic example of convergent evolution in action. Although their last common ancestor did not have wings, birds and bats do, and are capable of powered flight. The wings are similar in construction, due to the physical constraints imposed upon wing shape. Similarity can also be explained by shared ancestry, as evolution can only work with what is already there - thus wings were modified from limbs, as evidenced by their bone structure.

Similarity may also result if organisms occupy a similar ecological niches - that is, a distinctive way of life A classic comparison is between the marsupial fauna of Australia and the placental mammals of the Old World. The two lineages are clades - that is, they each share a common ancestor that belongs to their own group, and are more closely related to one another than to any other clade - but very similar forms evolved in each isolated population Many "body plans", for instance sabre-toothed cats and flying squirrels evolved independently in both populations. Convergence also occurs in cultural evolution.

Traits arising through convergent evolution are termed analogous structures, in contrast to homologous structures, which have a common origin. Bat and pterodactyl wings are an example of analogous structures, while the bat wing is homologous to human and other mammal forearms, sharing an ancestral state despite serving different functions. Similarity due to convergent evolution, but independent origins is called Homoplasy. Similarity due to the common ancestor is called Homology and not due to convergent evolution.

The opposite of convergent evolution is divergent evolution, whereby related species evolve different traits. On a molecular level, this can happen due to random mutation unrelated to adaptive changes; see long branch attraction. Convergent evolution is similar to, but distinguishable from, the phenomena of evolutionary relay and parallel evolution. Evolutionary relay describes how independent species acquire similar characteristics through their evolution in similar ecosystems, at different times: for example the dorsal fins of extinct ichthyosaurs and sharks. Parallel evolution occurs when two independent species evolve together at the same time in the same ecospace and acquire similar characteristics - for instance extinct browsing-horses and paleotheres.

Significance

The degree to which convergence affects the products of evolution is the subject of a popular controversy. In his book Wonderful Life, Steven Jay Gould argues that if the tape of life were re-wound and played back, life would have taken a very different course Simon Conway Morris counters this argument by arguing that convergence is a dominant force in evolution, and that since the same environmental and physical constraints act on all life, there is an "optimum" body plan which life will inevitably evolve towards, with evolution bound to stumble upon intelligence - a trait of primates, crows and dolphins - at some point Convergence is difficult to quantify, so there is no way to objectively resolve this argument.

Examples

One of the most famous examples of convergent evolution is the camera eye of cephalopods (e.g. squid) and vertebrates (e.g. mammals). Their last common ancestor had at most a very simple photoreceptive spot, but a range of processes led to the progressive refinement of this structure to the advanced camera eye - with one subtle difference; the cephalopod eye is "wired" in the opposite direction, with blood and nerve vessels entering from the back of the retina, rather than the front as in vertebrates The similarity of the structures in other respects, despite the complex nature of the organ, illustrates how there are some biological challenges (vision) that have an optimal solution. However, evolutionary history can act as a constraint, and most arthropods (eg insects) have a different form of eye, the compound eye. This is inferior to the camera eye in many respects, not least size and resolution. Arthropods have failed to evolve a camera eye as there is no way of getting there without a decrease in vision quality, and evolution has no capacity of foresight or advance planning.

LIST OF EXAMPLES OF CONVERGENT EVOLUTION

Convergent evolution - the evolution of similar traits in unrelated lineages - is rife in nature, as given below.

In Animals

Mammals

- Members of the two clades Australosphenida and Theria evolved tribosphenic molars independently.
- The marsupial Thylacine - Tasmanian Wolf, had many resemblances to the placental Canids.
- Several mammal groups have independently evolved prickly protrusions of the skin – echidnas (monotremes), the insectivorous hedgehogs, some tenrecs (a diverse group of shrew-like Madagascan mammals), Old World porcupines (rodents) and New World porcupines (another biological family of rodents). In this case, because the two groups of porcupines are closely related, they would be considered to be examples of parallel evolution; however, neither echidnas, nor hedgehogs, nor tenrecs

are close relatives of the Rodentia. In fact, the last common ancestor of all of these groups was a contemporary of the dinosaurs.

- Cat-like sabre-toothed predators evolved in three distinct lineages of mammals – sabre-toothed cats, Nimravids ("false" sabre-tooths), and the marsupial "lion" Thylacosmilus. Gorgonopsids and creodonts also developed long canine teeth, but with no other particular physical similarities.
- A number of mammals have developed powerful fore claws and long, sticky tongues that allow them to open the homes of social insects (e.g., ants and termites) and consume them (myrmecophagy). These include the four species of anteater, more than a dozen armadillos, eight species of pangolin (plus fossil species), the African aardvark, one echidna (an egg-laying monotreme), the enigmatic Fruitafossor, the singular Australian marsupial known as the numbat, the aberrant Aardwolf, and possibly also the Sloth Bear of South Asia, all not related.
- Koalas of Australasia have evolved fingerprints, very similar to those of humans.
- The Australian honey possums acquired a long tongue for taking nectar from flowers, a structure similar to that of butterflies, some moths, and hummingbirds, and used to accomplish the very same task.
- Marsupial Sugar Glider and Squirrel Glider of Australian are like the placental Flying Squirrel.
- The North American kangaroo rat, Australian hopping mice, and North African and Asian jerboa have developed convergent adaptations for hot desert environments; these include a small rounded body shape with very large hind legs and long thin tails, a characteristic bipedal hop, and nocturnal, burrowing and seed-eating behaviours. These rodent groups fill similar niches in their respective ecosystems.

- Opossums have evolved an opposable thumb, a feature which is also commonly found in the non-related primates. Marsupial mole has many resemblances to the placental Mole.
- Marsupial Mulgara - mouse has many resemblances to the placental mouse. Planigale has many resemblances to the Deer Mouse. Marsupial Tasmanian Devil has many resemblances to the placental Badger. Kangaroo has many resemblances to the Patagonian Cavy. The Marsupial lion had retractable claws, the same way the placental felines - cats do today. Microbats, toothed whales and shrews developed sonar-like echolocation systems used for navigation and for locating prey. Both the aye-aye lemur and the striped possum have an elongated finger used to get bugs from trees. There are no woodpeckers in Madagascar or Australia where the species evolved, so the supply of bugs in trees was large.
- Castorocauda and beaver both have webbed feet and a flattened tail, but are not related. Prehensile tails came in to a number of unrelated species New World monkeys'kinkajous, porcupines, tree-anteaters, marsupial opossums, and the salamander Bolitoglossa pangolins, treerats, skinks and chameleons. Pig form, large-headed, pig-snouted and hoofs are independent in true pigs in Eurasia and Peccary and Enteledonts.
- Plankton feeding filters, baleen: Whale sharks and baleen whales, like the humpback and blue whale independent have very sophisticated ways of sifting plankton from marine waters. River dolphins, there are five species of river/freshwater dolphins, they are not related.
- DNA study has shown that echolocation in two types of bats: megachiroptera and microchiroptera came about independently.
- Platypus have what looks like a bird's Beak (hence its scientific name Ornithorhynchus), but is a primitive mammal.

Dinosaurs

- Ornithischian (bird-hipped) dinosaurs had a pelvis shape similar to that of birds, or avian dinosaurs, which evolved from saurischian (lizard-hipped) dinosaurs.
- Yhe Heterodontosauridae evolved a tibiotarsus which is also found in modern birds. These groups aren't closely related.
- Ankylosaurs and glyptodont mammals both had spiked tails.
- Horned snouts independently is on non=related dinosaurs like ceratopsians and Triceratops, also rhinos and the brontotheres of the Cenozoic.
- Billed snouts on the duck-billed dinosaurs hadrosaurs strikingly convergent with ducks and duck-billed platypus.
- Ichthyosaurs a marine reptile of the Mesozoic era looked strikingly like porpoises.
- Beaks are independent in ceratopsian dinosaurs like Triceratops, birds and marine mollusks like squid and octopus.

Avian

- The Little Auk of the north Atlantic (Charadriiformes) and the diving-petrels of the southern oceans (Procellariiformes) are remarkably similar in appearance and habits.
- Penguins in the Southern Hemisphere evolved similarly to flightless wing-propelled diving auks in the Northern Hemisphere: the Atlantic Great Auk and the Pacific mancallines.
- Vultures are a result of convergent evolution: both Old World vultures and New World vultures eat carrion, but Old World vultures are in the eagle and hawk family (Accipitridae) and use mainly eyesight for discovering food; the New World vultures are of obscure ancestry, and some use the sense of smell as well as sight in hunting.

Birds of both families are very big, search for food by soaring, circle over sighted carrion, flock in trees, and have unfeathered heads and necks.

- Hummingbirds resemble sunbirds. The former live in the Americas and belong to an order or superorder including the swifts, while the latter live in Africa and Asia and are a family in the order Passeriformes.
- Certain longclaws (Macronyx) and meadowlarks (Sturnella) have essentially the same striking plumage pattern. The former inhabit Africa and the latter the Americas, and they belong to different lineages of Passerida. While they are ecologically quite similar, no satisfying explanation exists for the convergent plumage; it is best explained by sheer chance.
- Resemblances between swifts and swallows is due to convergent evolution.
- Downy Woodpecker and Hairy Woodpecker look the same, but are convergent evolution.
- Many birds of Australia, like wrens and robins, look like northern hemisphere birds but are not related.
- Oilbird like microbats and toothed whales developed sonar-like echolocation systems used for locating prey.
- The brain structure, forebrain, of hummingbirds, songbirds, and parrots responsible for vocal learning (not by instinct) is very identical. These types of birds are not related.

Reptiles

- The thorny devil (Moloch horridus) is similar in diet and activity patterns to the Texas horned lizard (Phrynosoma cornutum), although the two are not particularly closely related.
- Modern Crocodilians resemble prehistoric phytosaurs, champsosaurs, certain labyrinthodont amphibians, and perhaps even the early whale Ambulocetus. The

resemblance between the crocodilians and phytosaurs in particular is quite striking; even to the point of having evolved the graduation between narrow- and broad-snouted forms, due to differences in diet between particuler species in both groups.

- The body shape of the prehistoric fish-like reptile Ophthalmosaurus is similar to those of other ichthyosaurians, dolphins (aquatic mammals), and tuna (scombrid fish).
- Death Adders strongly resemble true vipers, but are elapids.
- The glass snake is actually a lizard but is mistaken as a snake.
- Large Tegu lizards of South America have converged in form and ecology with monitor lizards, which are not present in the Americas.
- Legless lizard-Pygopodidae are snake like lizard are much like true snakes.
- Mosasaurs of the Mesozoic era are like whales but are closely related to living monitor lizards and the Komodo Dragon.
- Anolis lizards are one of the best examples of both adaptive radiation and convergent evolution.

Fish

- Goby dorsal fin liked the lumpsuckers, yet not are related.
- Sandlance fish and chameleons have independent eye movements and focusing by use of the cornea.
- Cichlids of South America and the "sunfish" of North America are strikingly similar in morphology, ecology and behavior.
- The Peacock Bass and Largemouth Bass are excellent examples.
- The Antifreeze protein of fish in the arctic and Antarctic, came about independently.

- Eel form are independent in the North American brook lamprey, neotropical eels, and the African spiny eel.
- Stickleback fish, there is widespread Pconvergent evolution in Sticklebacks.

Amphibians

- Plethodontid salamanders and Chameleons have evolved a harpoon-like tongue to catch insects.
- Two lineages of frogs among the Neobatrachia are due for convergent evolution.
- The Neotropical poison dart frog and the Mantella of Madagascar have independently developed similar mechanisms for obtaining alkaloids from a diet of mites and storing the toxic chemicals in skin glands. They have also independently evolved similar bright skin colors that warn predators of their toxicity (by the opposite of crypsis, namely aposematism).

Arthropods

- Assassin spiders comprise two lineages that evolved independently. They have very long necks and fangs proportionately larger than those of any other spider, and they hunt other spiders by snagging them from a distance.
- The smelling organs of the terrestrial coconut crab are similar to those of insects.
- Silk: Spiders, silk moths, larval caddis flies, and the weaver ant all produce silken threads.
- The praying mantis body type – raptorial forelimb, prehensile neck, and extraordinary snatching speed - has evolved not only in mantid insects but also independently in neuropteran insects Mantispidae.
- Agriculture some kinds of ants, termites, and ambrosia beetles have for a long time cultivated and tend fungi for food. These insects sow, fertilize, and weed their crops. A damselfish also takes care of red algae carpets on its

piece of reef; the damselfish actively weeds out invading species of algae by nipping out the newcomer.

Molluscs

- The brachiopods (non-molluscs) and bivalve molluscs have very similar shells.
- There are limpet-like forms in several lines of gastropods: "true" limpets, pulmonate siphonariid limpets and several lineages of pulmonate freshwater limpets.
- Cuttlefish show similarities between cephalopod (nautili, octopods and squid) and vertebrate (Mammalia...) eyes.
- Swim bladders – Buoyant badders at independent in fishes, female octopus and jellyfish like the Portuguese Man o' War.
- Clamlike shells – Phylum Mollusca like clams and oysters and the Phylum Brachiopoda like brachiopods and lampshells, independent have invented paired shells for protection. The anatomy of their soft body parts is so dissimilar, however, that they are regarded as separate, independent phyla. Biologists think that clams are more closely related to earthworms than they are to brachiopods.
- Jet propulsion of squids and scallops, although both mollusks have independent very different ways of squeezing water through their bodies to power their movement through a fluid. Dragonfly larvae in the aquatic stage, use an anal jet to propel them. Jellyfish have had jet propulsion for a long time.
- The notochords in chordates are like the stomochords in hemichordates.
- Elvis taxon in the fossil record developed a similar morphology through convergent evolution.
- *Venomous sting:* To inject poison with a hypodermic needle, a sharppointed tube, has shown up independently 10+ times: jellyfish, spiders, scorpions, centipedes, various insects, cone shell, snakes, stingrays, stonefish, the male duckbill platypus, and stinging nettles plant.

- *Oxygenate blood:* Vertebrates use iron to bind to oxygen for transit through the blood system. Crustaceans and many mollusks use copper to bind oxygen in their blood system instead.
- *Bioluminescence:* A symbiotic partnerships with light-emitting bacteria developed many times independently in deep-sea fish, jellyfish, and in fireflies and glow worms.
- *Parthenogenesis:* Some lizards and insects have independent the capacity for females to produce live young from unfertilized eggs. Some species are entirely female.

In Plants

- Leaves have evolved multiple times - see Evolutionary history of plants.
- Prickles, thorns and spines are all modified plant tissues that have evolved to prevent or limit herbivory, these structures have evolved independently a number of times.
- *Hallucinogenic toxins:* Plants as diverse as the peyotyl cactus and the ayahuasca vine produce the same form of chemical toxin to deter predators.
- The aerial rootlets found in ivy (Hedera) are similar to those of the climbing hydrangea (Hydrangea petiolaris) and some other vines. These rootlets are not derived from a common ancestor but have the same function of clinging to whatever support is available.
- Flowering plants (Delphinium, Aerangis, Tropaeolum and others) from different regions form tube-like spur which contains nectar (that's why insect from one place sometimes can feed on plant from other which has such structure like the flower which is the traditional source of food for the animal).
- Both some dicots (Anemone) and monocots (Trillium) in inhospitable environments are able to form underground organs such as corms, bulbs and rhizomes for reserving of nutrition and water till the conditions become better.

- *Insectivorous plants:* Nitrogen-deficient plants have in at least 7 distinct times become insectivorous, like: flypaper traps/sundew, spring traps-Venus fly trap, and pitcher traps in order to capture and digest insects to obtain scarce nitrogen.
- Similar-looking rosette succulents have arisen separately among plants in the families Asphodelaceae (formerly Liliaceae) and Crassulaceae.
- The Orchids, the Birthwort family and Stylidiaceae have evolved independently the specific organ known as gynostemium, more popular as column.
- The Euphorbia of deserts in Africa and southern Asia, and the Cactaceae of the New World deserts have similar modifications.
- *Sunflower:* Some types of sunflowers and pevicallis are due to convergent evolution.

INDEX

F

G

H

❑❑❑